*PLANETS & PERCEPTION*

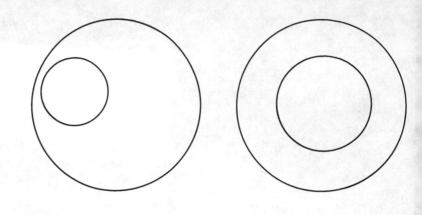

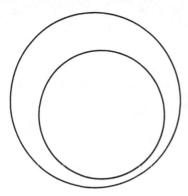

WILLIAM SHEEHAN

# *Planets &*

# *Perception*

*TELESCOPIC VIEWS*

*AND INTERPRETATIONS,*

*1609–1909*

*THE UNIVERSITY*

*OF ARIZONA PRESS*

*TUCSON*

The University of Arizona Press
www.uapress.arizona.edu

Printed in the United States of America
19  18  17  16  15  14    6  5  4  3  2

ISBN-13: 978-0-8165-1059-7 (cloth)
ISBN-13: 978-0-8165-3164-6 (paper)

Cover images: Percival Lowell observing Venus from the Lowell
Observatory in 1914; Mars Valles Marineris, courtesy of NASA

Library of Congress Cataloging-in-Publication Data
Sheehan, William, 1954–
    Planets and perception : telescopic views and interpretations,
1609–1909 / Wiliam Sheehan.
        p.      cm.
    Bibliography: p.
    Includes index.
    ISBN 0-8165-1059-8 (alk. paper)
    1. Planets—Observations. 2. Perception. I. Title. II. Title:
Planets and perception
QB601.S54 1988
522'.2'0903—dc19                                          88-20501
                                                              CIP


♾ This paper meets the requirements of ANSI/NISO Z39.48-
1992 (Permanence of Paper).

*To Arthur A. Hoag*

*and in memory of*

*William Graves Hoyt*

# CONTENTS

# ILLUSTRATIONS

AN AUTHOR knows few pleasures
greater than that of giving recognition to those who have lent
their assistance to his labors. So it is here. Arthur A. Hoag, the
former director of the Lowell Observatory, was most gracious
in making available the archives of that observatory and in
allowing me to use its 24-inch Clark refractor. He later read
the book in manuscript form and annotated it thoughtfully
and constructively. I also received much help from the Lowell
Observatory's resident historian, the late William Graves
Hoyt. Mary Lou Evans and Stuart Jones of the Lowell Observatory were very helpful in providing illustrations, and Jay
Gallagher, the present director of Lowell, was kind enough to
grant them time to do so.

Judith Bausch, the librarian of the Yerkes Observatory, and
Richard Dreiser of its photographic department were most
helpful in locating and making available material from that
observatory's archives and kindly lent their assistance during
my visit to the Yerkes Observatory. In particular, I would like
to express my appreciation to them for calling to my attention
the drawings of Mars by E. E. Barnard that are published
here.

Among others who either commented on the manuscript in
whole or in part or who exerted a direct influence on my
thinking by sharing with me their expertise were Clyde W.
Tombaugh, the discoverer of Pluto and now Professor Emer-
itus at New Mexico State University, Dale P. Cruikshank of
the University of Hawaii, Michael J. Crowe of the University
of Notre Dame, and Michael Armstrong of the University
of Illinois at Champaign. Rodger W. Gordon and Terence
Dickinson gave me their views concerning Martian crater
observations and problems of chromatic aberration in
visual planetary astronomy, respectively. Though most of the
original materials I drew upon were in English, my researches did carry me to a lesser extent into French, German,
Italian, and Latin sources. For their help in providing translations of some of these, I have Michael Armstrong again to
thank, as well as my brother, Bernard J. Sheehan. *xi*

The following individuals and firms granted me permission to reprint from works over which they hold copyright: Doubleday & Company, Inc., for permission to reprint from *Discoveries and Opinions of Galileo,* translated and with an introduction and notes by Stillman Drake, copyright © 1957 by Stillman Drake; Harcourt Brace Jovanovich, Inc., for permission to reprint from Edward Rosen, translator, *Kepler's Conversation with Galileo's Sidereal Messenger,* copyright © 1965 by Johnson Reprint Corporation; Allen and Unwin, Inc., for permission to reprint from Antonie Pannekoek, *A History of Astronomy,* copyright © 1961 by Allen and Unwin, specifically his translation of passages from Camille Flammarion's *La Planète Mars,* 1892; Dr. Patrick Moore for permission to quote from his translations of E. M. Antoniadi's *The Planet Mars* and *The Planet Mercury,* copyright © 1974 and 1975 by Dr. Patrick Moore; Macmillan Publishing Company for permission to reprint from Abbott Lawrence Lowell, *The Biography of Percival Lowell,* copyright © 1935 by The Macmillan Company; Houghton Mifflin Company for permission to reprint from Ferris Greenslet, *The Lowells and Their Seven Worlds,* copyright © 1946 by Ferris Greenslet and copyright © renewed 1973 by Magdalena Greenslet Finley; Princeton University Press for permission to reprint from E. H. Gombrich, *Art and Illusion: A Study in the Psychology of Pictorial Representation,* No. 5 in the A. W. Mellon Lectures in the Fine Arts, Bollingen Series XXXV, copyright © 1960 by The Trustees of the National Gallery of Art.

Finally, Michael Conley has encouraged my work on the planets, my parents, Bernard L. and Joyce Sheehan, have been supportive as ever, and my wife, Deborah, has helped me in more ways than I can name.

To these, and to others in lesser ways, much of what is of merit in these pages is due; any errors that remain, of course, are my responsibility.

William Sheehan
St. Paul, Minnesota
June 5, 1987

PLANETS & PERCEPTION

## CHAPTER 1

# Victorians, Apes, and Martians

*And Mars alone holds the skies.*
          —*Lucan,* Pharsalia

ON MARS HILL, the mesa rising above the city of Flagstaff, Arizona, stands the dome of the 24-inch refractor Percival Lowell once used for his planetary observations. On Lowell's mausoleum, a stained-glass bubble of sky blue not far from the dome itself, an inscription reads in part, "To see into the beyond requires purity . . . and the securing it makes him perforce a hermit from his kind."[1] In the many years since Lowell built his observatory, it has been encroached upon by the growing city, and the streetlights shining through the trees below outshine the stars' pale lamps. Those concerned with viewing the stars must seek more distant, darker skies; for them the separation Lowell wrote about remains a prerequisite. But to the planetary observer the lights below offer a less serious impediment to seeing the lights above than might be supposed. They detract but little from the view—aside, perhaps, from what they may steal from one's aesthetic appreciation. The observer's concern is with stillness rather than with darkness, and to obtain stillness he may even observe in broad daylight. The twinkling of the stars, so warm and comforting to the rest of humankind, is fatal to his aspiration: that twinkling proves that their light must pass through a blanket of air, insulating the Earth from the cold and empty barrenness of space. In constant motion, the ebb and flow of the atmospheric tides—for ocean it is—smear and distort detail.

Sunset. The San Francisco Peaks are fringed briefly with pink where the splendor has not yet entirely subsided. Then

*1*

they too grow dark, and a hush settles over the fragile pine forests of the Coconino Plateau of northern Arizona. One by one the stars appear—the seemingly eternal and unchanging stars following close on the perishing beauty of the fading sunset.

I stroll up to the dome, where the shaft of the telescope Lowell once used rises almost perpendicularly into the air and hangs suspended as if weightless. The chair in which he sat invokes his ghostly presence in the dull red glow of the night-lights. The tapering lead-grey cylinder points the eye upward to the ceiling, where graceful ribs support the dome—the cathedral to a new kind of faith, perhaps. Lest it be judged inferior by comparison with the cathedrals of earlier ages, magnificent with their vast stretches of weightless stone and fan vaults, the scene on which its shutter opens manually by ropes and pulleys is far more beautiful than any window of stained glass. In a swath of sky a planet, jewel-like, glistens.

My hand directs the telescope toward westering Mars and swivels the dustcover from the lens. With mounting anticipation I take my place in the observing chair and lean toward the eyepiece, within whose field swims the small salmon-colored disk I seek. Over it, intermittently, uncertain details dart and then are gone like doubtful visions. It is the fluctuating air that first presents then withholds them from my view. We who now know what lies beyond the shimmering veil may forget how tantalizing that orb once was. Astronomers no longer peer through telescopes as their primary mode of research. They now have far better eyes—those of spacecraft orbiting near the planets their grandfathers could view only remotely. They have obtained maps of the planetary surfaces exquisite in their detail. Some of the markings vouched for by the old observers have been relegated to the realm of the mythical. Yet while we may applaud the cumulative advance of science which, in Tennyson's words,

> reaches forth her arms
> To feel from world to world, and charms
> Her secret from the latest moon,[2]

it is not, perhaps, without a modicum of regret that we must leave the myths—some rather lovely in their way—behind.

IN THE AUTUMN of 1877 an Italian astronomer, Giovanni Virginio Schiaparelli, discovered the markings on Mars that he termed *canali*. In Italian this can mean "channels" as well as "canals," but it was as the latter that it was rendered into English. At first the *canali* looked like a class of natural features of the planet, though admittedly unlike anything on Earth. As Schiaparelli continued to observe, however, they came to appear more regular and geometric. With the completion of the Suez Canal in 1869, one of the great engineering feats of the century, it was not long before others, if not Schiaparelli himself, drew the obvious conclusion. The person who had translated *canali* into "canals" had been right, or so it seemed. The markings were apparently canals in fact as well as in name—true artifacts on the planet's surface.

Earlier in the century William Paley, in elaborating the age-old proof of the existence of God by the argument from design in nature, had written in his book *Natural Theology* (1802) that "it is only by the display of contrivance, that the existence, the agency, the wisdom of the Deity *could* be testified to his rational creatures." Percival Lowell would soon be writing about the beings that were presumed to lie behind the rational works appearing on the Martian surface: "Too small in body himself to show, it would only be when his doings had stamped themselves there that his existence could with certainty be known. Then and only then would he stand disclosed." Paley had written: "The marks of *design* are too strong to be gotten over. Design must have had a designer. That designer must have been a person. That person is God." Lowell would write of his Martians: "His mind would reveal him by his works—the signs left upon the world he had fashioned to his will. And this is what I mean by saying that through mind and mind alone we on Earth should first be cognizant of beings on Mars."[3] Where there was a watch, there had to have been a watchmaker; where there were canals, there had to have been canal builders. Both arguments were framed along the lines of a primitive, childlike tendency to construct artificialist explanations for natural phenomena. When children are asked how the Sun began, notes psychologist Jean Piaget, they almost always declare that humans made it; he deems it probable that this artificialist answer "is connected with a latent artificialism, an artificialist tendency of mind natural to children"[4]—and, we may add, not only to children.

Yet if the Martians were descended from Paley on one side, they could also claim descent from Darwin, whose theory of evolution was, in a sense, conceived as a rebuttal of Paley. Darwin made humankind's position in the biological scheme relative less than two decades before Schiaparelli's discovery of the Martian "canals." Man, the "wonder and the glory of the universe," could no longer so smugly view himself as having the sole privilege of this universal frame. He had surpassed in mind the other denizens of Earth, but this did not mean that he might not someday, or somewhere, be surpassed himself. The conclusion that had remained implicit in *The Origin of Species,* published in 1859, was made explicit by the time Darwin published *The Descent of Man* in 1871:

> The main conclusion arrived at in this work, namely, that man is descended from some lowly organised form, will, I regret to think, be highly distasteful to many. . . . We must, however, acknowledge, as it seems to me, that man with all his noble qualities, with sympathy which feels for the most debased, with benevolence which extends not only to other men but to the humblest living creature, with his god-like intellect which has penetrated into the movements and constitution of the solar system —with all these exalted powers—Man still bears in his bodily frame the indelible stamp of his lowly origin.[5]

The revolution of Darwinian thought extended far beyond biology, just as that of Copernican thought had extended far beyond astronomy. It gave rise, for instance, to the philosophy of Herbert Spencer and the sociological theory of Social Darwinism, which held industrial society up as a mirror of the biological "struggle for existence." Whereas Tennyson had recoiled from a nature "red in tooth and claw,"[6] the Social Darwinists found it gratifying to note that the capitalist system, however flawed, was fashioned along the lines of the natural order and thus had to be the best possible system.[7] When Mars came to be more closely scrutinized with the large telescopes being built especially in the United States toward the end of the nineteenth century (many of them endowed by such millionaire capitalists as James Lick, Charles Yerkes, and

Andrew Carnegie) and when the planet's light was analyzed
by the spectroscope for evidence of air and water, it seemed to
be, not absolutely hostile, but certainly harsher in its conditions than the Earth. Who knew to what possibly higher summit of the organic scale than man the "struggle for existence"
on Mars—regarded by some as an older world than our own—
might have led?

With this we come to the Martian theories of Percival Lowell, himself a millionaire and a Social Darwinist. Lowell was
responsible for the idea that the markings Schiaparelli had discovered on Mars were actual canals that had been dug by the
Martians to transfer water from the polar caps, where most of
the planet's limited supply was stored, to the thirsty deserts.
The whole planet was, as he saw it, dying a slow death by
desiccation. Lowell tied his Martian theories to a broad evolutionary scheme of planetary development which lent greater
plausibility to them for the uncritical reader. As he saw it, the
Earth's fate was to be the same as that of Mars. A planetwide
desert was the inevitable result of the evolution of worlds.
What Mars now was, therefore, the Earth would oneday be.
Mars occupied, he wrote, "earthwise in some sort the post of
prophet, . . . foretelling our future. From a scrutiny of Mars,
coming events cast not their shadow, but their light" before
them. "It is the planet's size that fits it thus for the role of
seer," he wrote. "Its smaller bulk has caused it to age quicker
than our Earth, and in consequence, it has long since passed
through that stage of its planetary career which the Earth at
present is experiencing, and has advanced to a further one to
which in time the Earth must come."[8] And he added, in apocalyptic refrain, "The outcome is doubtless yet far off, but it is
fatalistically sure as that tomorrow's Sun will rise, unless some
other catastrophe anticipate the end."[9]

Lowell, whose own mind was thoroughly captivated by the
idea of intelligent beings on Mars, tended to underestimate
the allure of that idea for others. He saw himself as a per-
secuted pioneer like Darwin, struggling against conservative
resistance to one of the master ideas of the age. Yet whereas
Darwin had rightly feared that others would reject the idea
that man was descended from lower forms, Lowell did not, in
fact, face a similar rejection when he offered the prospect of
man's being superseded by higher forms. Though he claimed

that "man is so profoundly jealous of peers that he looks askance at anything resembling himself, like a savage at his own reflection in a looking glass,"[10] he forgot that, to the contrary, humans had always fashioned beings resembling but superior to themselves—and had filled the sky with them. With his Martians, he simply brought the gods in by the back door. If one wishes, one may explain the universal tendency to look up to gods of one sort or other as grounded in the child's awed view of his or her parents;[11] but in any case, there can be no dispute about the tendency.

Since Lowell's day we have seen a recrudescence of this seemingly irrepressible need of the human psyche in the form of the wide credence given to UFOs in the postwar era, and in the current quest for extraterrestrial civilizations, whose rhetoric is also to a significant degree that of a quasi-religious movement based as much on emotion as on scientific plausibility.

The question of extraterrestrial life undoubtedly falls within the legitimate domain of science. As Bernard de Fontenelle, a seventeenth-century predecessor of Lowell, wrote, "What can more concern us, than to know how this world which we inhabit, is made; and whether there be any other worlds like it, which are also inhabited as this is?"[12] On the other hand, few scientific questions have been more frequently subjected to such unscientific treatment. In particular, the controversy over Mars at the turn of the century had some harmful effects on planetary science, leading to an ostracism of planetary astronomers such as G. P. Kuiper from the rest of the professional astronomical community long after.[13] Clearly, much of the blame lies with the speculative writings of such men as Lowell. In the end, though, for all their speculations, astronomers could discover little about the Martians themselves; the discussion came to focus, rather, on the question of how sound the observations were on which the astounding conclusions rested—and rested precariously, in the eyes of those who were disinterested enough to properly assess them.

Indeed, the planetary observer faced daunting obstacles as he sat at his telescope trying to make out what lay at the other end of the column of light he had captured. In the case of Mars, the disk was small, the features more or less indistinct. Moreover, the advantage that was optimistically expected when larger telescopes came to be pointed toward it was to

some extent—some thought altogether—offset by the deleterious effects of the Earth's own atmosphere, through which
light from Mars had necessarily to follow its troubled way.
The larger the telescope, the more adversely the atmosphere
seemed to enter in. The explanation for this is statistical. For
the planet's image to be sharply defined, the roiling air above
the telescope had to settle down at least for a moment to allow
the undisturbed passage of the planet's light. The wider the
column of air in which this settling had to exist, the less
frequently such moments of relative clarity could be expected
to occur. With a very large telescope—Lord Rosse's 6-foot reflector, erected in the middle of the nineteenth century, for
instance—they might occur very infrequently indeed, though
they would still not be altogether absent. Thus G. Johnstone
Stoney, the Irish astronomer who used the great Rosse reflector regularly for a time, noted that "if the night is good, there
will be moments now and then when the atmospheric disturbance will abruptly seem to cease for a fraction of a second."[14]
The observer's ability to remember what was vouchsafed to
him during the momentary revelations no doubt played an
important role in the accuracy of his representation, as did his
ability to draw what he had seen. E. Walter Maunder, an
English astronomer, wrote:

> The drawing cannot be made at the very instant that the
> original is being viewed. The degree, therefore, to which
> some point attracts special attention will have much to do
> with the relative proportions given to it. Individual dis-
> cordances are often considerable, and "changes on Mars"
> have been advertised in the most reckless fashion, on the
> faith of a single drawing, quite neglectful of this source of
> error even with the best artists.[15]

Or, as Sir Winston Churchill has written of the role of memory in artistic representation generally,

> It would be interesting if some real authority investigated
> carefully the part which memory plays in painting.
> . . . The canvas receives a message dispatched usually only
> a few seconds before from the natural object. But it has
> come through a post office *en route*. It has been transmit-
> ted in code. It has been turned from light into paint. It

reaches the canvas a cryptogram. Not until it has been placed in its correct relation to everything else that is on the canvas can it be deciphered, is its meaning clear, is it translated once again from mere pigment into light. And the light this time is not of Nature but of Art.[16]

The question of what happens when the message comes through the "post office"—the mind—will become our central theme. Numerous, indeed, as are the factors which we must examine as influences on visual planetary observation—of which the optical and atmospheric are the most obvious—in the end the main factor appears to be the psychological one. It is therefore only fair to warn the reader in advance that sources of illusion will be quite as much to our purpose as the known facts about planetary surfaces. For that matter, the line between the two was seldom clearly discernible to the participants themselves.

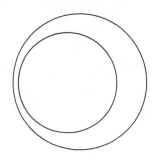

CHAPTER 2

# The Face in the Moon's Disk

THE EXPLORATION OF the planets may conveniently be regarded as falling into three distinct periods. The most recent is scarcely twenty years old. It is that in which humankind, by means of spacecraft, has begun direct planetary exploration, an adventure that cannot help but have, indeed has already had, the most profound impact on human thought. But it is too recently underway to fall properly within the historian's perspective, being still more present than past.

In contrast, the beginning of the first of the three periods, that of naked-eye observation, took place in remote antiquity, when the planets were first recognized for what their name implies them to be—planetae, or wanderers, which indeed they are against the background of the other stars. In those remote days, the planets were merely points of light. The main problem connected with them was that of explaining their motion, and this problem was finally solved in 1609 by Johannes Kepler, a onetime assistant of the greatest of the exclusively naked-eye observers of the heavens, Tycho Brahe. Through a process by turns painful and exhilarating, Kepler was able to use Tycho's hoard of careful observations as the basis for overcoming at last the two-thousand-year-old preju-dice in favor of circular orbits (or circles-on-circles, the noto-rious epicycles), proving that the actual paths of the planets are ellipses, with the Sun lying in one focus and the other focus being empty—the first of the three Keplerian laws of planetary motion. For later reference, the other two are:

Law II. The radius vector, that is, the line connecting the planet to the Sun, sweeps out equal areas in equal times.

Law III. The squares of the periods in which the planets describe their orbits are proportional to the cubes of their mean distances from the Sun.

It turns out that the difference between the elliptic orbits in which the planets actually travel and perfectly circular orbits is slight; in fact, it is just barely enough to have been detected from the most accurate observations that can be made with the naked eye, such as those of Tycho.

The formulation of Kepler's laws of planetary motion was the greatest achievement of the naked-eye period of planetary exploration, crowning two thousand years of observation and calculation, throughout which the planets could hardly be considered otherwise than as mere mathematical points describing paths in some abstract geometrical space. Yet already in Kepler's time a new instrument had become available that was to completely revolutionize astronomy and magnify those points into worlds. In 1609—the same year that Kepler published the first two of his three laws—Galileo, Thomas Harriot, Simon Marius, and a few others turned the first telescopes to the sky. The planets began to emerge as other worlds—if not, indeed, as other Earths—as they took shape as spheres with surfaces of rock, ocean, or cloud. The curtain had risen on a second period of planetary astronomy—fittingly, in the same year in which, in a sense, it had fallen on the first.

GALILEO STOOD head and shoulders above the rest of the early telescopic observers. His curiosity was, naturally enough, first piqued by the Moon. As the closest celestial body to the Earth, the Moon shows even to the naked eye the large patches on its surface that make up the face of the "Man in the Moon." The Pythagoreans are said to have believed that these patches formed a mirror reflection of the Earth's own oceans and lands. Though the attribution to the Pythagoreans is doubtful, the idea was unquestionably an early one. It appears in *De Facie in Orbe Lunae* (or, in what would be a better translation of the Greek, *De Facie quae in Orbe Lunae Apparet*—"On the Face Which Appears in the Moon's Disk"), written in the second century A.D. by Plutarch of Chaeronea. The work is in the form of a dialogue, and one of the characters mentions the opinion of one Clearchus that "the face, as we call it, is made up of images of the great ocean mirrored in the Moon."[1] Another participant in the dialogue attributes to the same Clearchus the belief that "the Moon is a mixture of air and fire," while a third comes forward to denounce both of these opinions. This participant—who is, according to Harold Cherniss, to be identified with Plutarch himself[2]—says that "it is a slap in the face to the Moon when they fill her with smuts and blacks, addressing her in one breath as Artemis and Athena, and in the very same describing a caked compound of murky air and charcoal fire, with no kindling or light of its own, a nondescript body smoking and charred."[3]

Certainly the idea that the Moon's "face" was a reflection of the Earth persisted for a long time, at least in the popular mind. Kepler's patron, Rudolf II, argued shortly before Galileo began his telescopic work that "in particular Italy with its two adjacent islands seemed to him to be distinctly outlined."[4] As late as the nineteenth century a visitor to Paris from Ispahan, on being shown the Moon through a telescope, told Alexander von Humboldt that in Persia it was still widely believed that "what we see in the Moon is ourselves; it is a map of Earth."[5]

Others contended that the lunar features were not merely reflections but actual oceans and lands. Thus Leonardo da Vinci regarded the bright areas as seas, the dark spots as continents. So did William Gilbert, best known as the author of *De Magnete,* who in 1600—on the eve of the invention of the telescope—drew up the first lunar chart and the only one which, so far as we know, was ever made in the naked-eye era.[6] Kepler, too, for a time concurred in thinking of the dark spots as lands and the light areas as seas.

There matters stood until November 1609, when Galileo first pointed his telescope at the Moon. He wrote of his observations in a little book, *Sidereus Nuncius* (Starry Messenger), published early in the following year:

On the fourth or fifth day after new moon, when the moon is seen with brilliant horns . . . many luminous excrescences extend beyond the boundary into the darkened portion, while on the other hand some dark patches invade the illuminated part. Moreover a great quantity of small blackish spots, entirely separated from the dark regions, are scattered almost all over the area illuminated by the sun with the exception only of that part which is occupied by the large and ancient spots. . . . There is a similar sight on earth about sunrise, when we behold the valleys not yet flooded with light though the mountains surrounding them are already ablaze with glowing splendor.[7]

Galileo's observations established that if there *were* lands and seas on the Moon, it was the dark areas (the "ancient spots") that had to be the seas (Fig. 2.1). He continued,

As to the large lunar spots, these are not seen to be broken in the above manner and full of cavities and prominences; rather, they are even and uniform, and brighter patches crop up only here and there. Hence if anyone wished to revive the old Pythagorean opinion that the moon is like another earth, its brighter part might very fitly represent the surface of the land and its darker region that of the water. I have never doubted that if our globe were seen from afar when flooded with sunlight, the land regions would appear brighter and the watery regions darker.[8]

Kepler read Galileo on this point as a revelation. In his *Conversation with Galileo's Starry Messenger,* written in April 1610, he recalled that his misconception had been based on an observation he had made on Mt. Schöckel in Styria. The river below had appeared brilliant at a time when all the surrounding land had appeared dark. But he now realized that this had been simply owing to reflection by the water of the bright sky

and that from directly overhead the river would have appeared dark. Moreover, in showing that sunrise came first to the bright areas, Galileo had proved that the dark areas were lower. Kepler at once declared, "the spots are seas, the light areas are lands."[9]

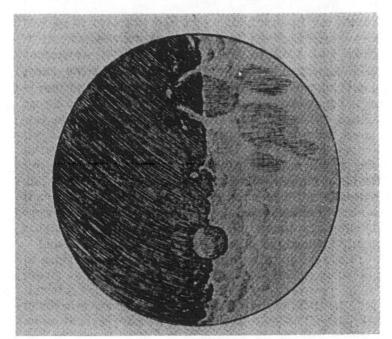

*Figure 2.1*
*One of Galileo's drawings of the*
*Moon, from the* Sidereus Nuncius
*(Starry messenger), of 1610. It*
*shows a large crater on the*
*terminator which he said he*
*"beheld . . . not without a certain*
*wonder." Though he professed to*
*have exercised a great deal of care*
*in executing the drawing, the*
*identification of the large crater*
*with any known feature is*
*uncertain.*

Kepler was not the only one to read the *Sidereus Nuncius*
as a revelation. England's Thomas Harriot had observed the
Moon with a telescope as early as July 1609—even before
Galileo. His first drawing, produced at that time, shows little
insight into the true nature of what he was seeing, about as
little, indeed, as was shown by his friend Sir William Lower
on that memorable occasion when, in describing the Moon's
appearance in his telescope, he said that it looked like a tart his
cook had made— "here some bright stuff, there some dark,
and so confusedly all over." Harriot's second drawing, made a
year later—after he had read the *Sidereus Nuncius*—is much
better and shows unmistakably the influence of Galileo's published drawings.[10] This only goes to show that true seeing is
not with the eye but with the mind—or, as Sir John Herschel
said, that "an object is frequently not seen *from not knowing
how to see it,* rather than from any defect in the organ of
vision."[11] Before he read the *Sidereus Nuncius,* Harriot had not
known *how* to see the Moon. Lower had had no more success
with the Pleiades, the star cluster in Taurus, in which Galileo
had reported many new stars. Lower wrote that he had observed "the seven stars also in Taurus, which before I always
rather believed to be seven than ever could number them.
Through my Cylinder I saw these also plainly and far asunder,
and more than seven too, but because I was prejudiced with
that number, I believed not mine own eyes nor was careful to
observe how many."[12]

Galileo was clearly far in advance of his contemporaries in
knowing *how* to see. Yet even *his* drawings look very crude by
today's standards. He records, for instance, a particularly large
crater at the terminator of the Moon (the line dividing day
from night) which simply does not correspond—in scale, at
any rate—to any recognized feature. There seems to be no
choice but to conclude, as Terrie F. Bloom has done, that the
size of the crater as Galileo records it only "preserves the
psychological impact of his interpretation."[13] Nevertheless, it is
at least interesting that Galileo in the *Sidereus Nuncius* makes
a specific point of the accuracy with which he made his depiction. He writes, "There is another thing which I must not
omit, for I beheld it not without a certain wonder; this is that
almost in the center of the moon there is a cavity larger than
all the rest, and perfectly round in shape. I have observed it

near both first and last quarter, and have tried to represent it as correctly as possible."[14]

The apparent discrepancy between the effort expended and the result only underscores the truth of a point made by Sir Ernst Gombrich: "To draw an unfamiliar sight presents greater difficulties than is usually realized."[15] The early attempts of observers to represent what they saw through their telescopes might, indeed, be compared to those of Renaissance artists faced with exotic animals such as whales, lions, and rhinoceroses. Some of the inaccuracies in the representations are amusing today, yet in some cases they were drawn by artists of consummate skill—Albrecht Dürer, for instance.

In other drawings Galileo showed the planet Saturn with what look like a pair of ears attached to its globe. Having discovered the four satellites of Jupiter not long before this, he concluded that the "ears" were two large satellites, writing to that effect to Giuliano de' Medici, "Behold! we have found the court of Jupiter, and now also for this old man two servants who help him walk and never leave his side."[16] Two years later, however, the supposed satellites had disappeared, seemingly without trace, and Galileo wrote in dismay, "Were the appearances indeed illusion or fraud with which the glasses have so long deceived me, as well as many others to whom I have shown them?"[17] Subsequently they reappeared, but in any case Galileo never fully grasped the nature of what he was seeing. Neither did a number of others who observed the planet (Fig. 2.2). The Frenchman Pierre Gassendi, in particular, failed to do so despite having assembled a fairly complete set of drawings. Christiaan Huygens later described the drawings of his predecessors (Fig. 2.2) as representing the planet under "a succession of . . . strange and marvelous forms . . . forms of such unusual appearance that they were considered by many as a mockery of the eyes."[18] This would not be the last such case in the history of visual planetary observation. Yet it was not that the eyes of the observers were deceiving them; it was rather that their minds did not know how to interpret the data with which they were being supplied.

R. L. Gregory has noted that "drawings do not represent how the object appears through a telescope at any one time. They are a synthesis of very many observations. They represent a belief in what it is 'really' like."[19] The synthesis Gregory

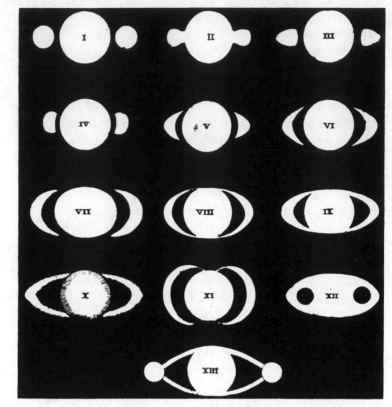

*Figure 2.2*
*Early telescopic views of Saturn*
*collected by Christiaan Huygens*
*and published in his* Systema
Saturnium, *of 1659. The views had*
*originally been drawn by Galileo*
*(I), Christoph Scheiner (II),*
*Giovanni Riccioli (III, VIII, IX),*
*Hevelius (IV–VII), Eustachio*
*Divini (X), Francesco Fontana*
*(XI, XIII), Pierre Gassendi (XII).*

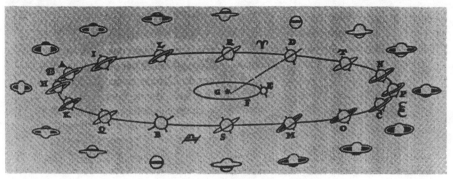

*Figure 2.3*
*Huygens's diagram from* Systema
Saturnium, *illustrating how
the hypothesis that Saturn is
surrounded by a "thin, flat ring,
nowhere touching," could account
for the various observed telescopic
appearances of the planet.*

speaks of usually includes not only one's own observations but those of others as well. In some cases an earlier representation may introduce a beneficial prejudice, but of course the influence may also work the other way. Precisely because it is so difficult to draw an unfamiliar sight for the first time, the drawings of later observers are, indeed, fairly likely to be fashioned as much after other drawings as "from life," so once an erroneous schema appears, it may be very difficult to get rid of it. We shall return to this again later when we consider the post-Schiaparellian drawings of Mars.

Only after Christiaan Huygens recognized in 1659 that Saturn is surrounded by "a thin, flat ring, nowhere touching," as he wrote in his *Systema Saturnium,* did the observations of Saturn finally fall into place (Fig. 2.3). The breakthrough was only partly one of telescopic performance—Huygens's telescope, with an object glass of 2⅓ inches and a magnification of 100, was, though the best of its day, not significantly better than Gassendi's. What Huygens had achieved was rather a breakthrough in perception. He had not seen a significantly better image, but he had discovered *how* to see it. Well did he deserve the paean that his father wrote for him, recognizing that he had made a discovery that would assure him of immortality:

> Glory coeval with the stars on high,
> To die only when heav'n itself shall die.[20]

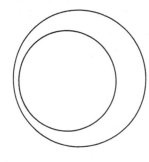

CHAPTER 3

# A Multitude
# of Earths

*We may mount from this dull Earth,
and viewing it from on high, consider
whether Nature has laid out all her cost
and finery upon this small speck of
Dirt. . . . We shall be less apt to admire
what this World calls great, shall nobly
despise those Trifles the generality of
Men set their Affections on, when we
know that there are a multitude of such
Earths inhabited and adorn'd as well as
our own.*
—Christiaan Huygens, The Celestial
Worlds Discovered

GALILEO HIMSELF did not assume that the difference between light and dark areas on the Moon was that between lands and seas. In writing to Giacomo Muti in 1616 he said, "I do not believe that the body of the Moon is composed of earth and water."[1] By the time he wrote his *Dialogue Concerning the Two Chief World Systems,* published in 1632, it is clear that he had not changed his mind. He wrote:

> If there were in nature only one way for two surfaces to be illuminated by the sun so that one appears lighter than the other, and that this were by having one made of land and the other of water, it would be necessary to say that the moon's surface was partly terrene and partly aqueous. But because there are more ways known to us that could produce the same effect, and perhaps others that we do not know of, I shall not make bold to affirm one rather than another to exist on the moon.[2]

It was unfair, then, of the poet Milton, who visited the aged and blind Galileo while he was under house arrest by the Inquisition at Arcetri, to have described him in *Paradise Lost* as one who

> views
> At Ev'ning, from the top of Fesole,
> Or in Valdarno, to descry new Lands,
> Rivers, or Mountains in her spotty Globe.[3]

Not only was Milton mistaken about the site of Galileo's observations—they were made neither at Fesole (near Florence) nor at Valdarno (or Valdagno, Venice's province), but at Padua. More important, he was mistaken about Galileo's motivations. Milton's elocution "views . . . to descry" implies purpose rather than accomplishment[4] and raises the important question of whether the purposeful seeking might not itself prove capable of investing the "spotty Globe" with "Lands, Rivers, or Mountains." Galileo, at least, was not guilty of this. Yet others soon were. Though the idea of the face in the Moon as a *literal* reflection of the Earth could no longer, after Galileo, be seriously countenanced, now that the telescope had revealed the Moon to be a world in its own right, the idea reappeared in modified form. For it proved only too tempting to fashion the Moon in the image of the Earth and to see on the Moon and the other planets counterparts of the terrestrial oceans and lands. Indeed, just as astronomers during the first period of planetary exploration had been dominated by the image of the circle, those of the second were to be dominated by that of the Earth, with its oceans and lands.

John Wilkins, bishop of Chester, whose *Discovery of a World in the Moone* was published in 1638, took as one of his basic propositions that "those spots and brighter parts which by our sight may be distinguished in the Moone, doe shew the difference betwixt the Sea and Land in that other World," from which he concluded that "'tis probable there may be inhabitants in this other World, but of what kinde they are is uncertaine."[5] The lunar map drawn by Langrenus in 1645 fixed the impression by giving watery names to the dark spots. When Hevelius two years later rejected Langrenus's scheme of nomenclature, which used the names of well-known personalities for various features (he had wanted to avoid arousing the jealousy of those who were left out, Hevelius said), he turned to a geographical nomenclature because he "knew not with what else to compare" the lunar features. Such nomenclature could not help but underscore the presumptive analogy to the Earth. Though Hevelius's geographical system proved clumsy in practice and did not survive—on the 1651 map of the Jesuit astronomer Giovanni Riccioli it was replaced by one more along the lines of Langrenus's[6]—such terms as "sea," "ocean," and "lake" remained firmly in place, and arguably had an influence on what other observers saw there. Indeed, referring to the later map of the Moon drawn by the seventeenth-century astronomer Giovanni Domenico Cassini, John Russell, who made a series of beautiful crayon drawings of the Moon in the next century, wrote that

> the Moon requires much attention to be well understood, being composed of so many parts, of different characters, so much similitude in each Class of form, and of such a variety in the minutiae composing those Forms and this difficulty also much considerably increasd by the various effects caused by the different situations of the Sun; that I am perswaded many considerable improvements may be made, in correctness of form in the spots, their situation and distinctness of parts; *perhaps it was too hastily concluded that the large dark parts upon the Moon's Face, were Seas;* I am apprehensive that, if the Engraver had been faithfull to his trust, this must have led that great Astronomer Cassini to represent these parts of one almost uniformly smooth, and u[n]varied effect, which upon a strict inspection will appear to be full of parts as various and nearly as multitudinous, as that portion of the Moon, which has generally been considerd to be Land. (Italics mine.)[7]

The tendency to represent parts as "uniformly smooth, and unvaried" will appear again when we come to consider the observations of the dark spots of Mars, which would also, for a time, be regarded as seas. But they too would show, when subjected to closer examination, "parts as various and nearly as multitudinous" as had been reported in those areas that had hitherto been regarded as lands.

ONE SOMETIMES GETS the impression that all Galileo had
to do was turn his telescope to the sky for discoveries to pour
from it as from a cornucopia, that the tour de force described
in the *Sidereus Nuncius* was made possible by the mere introduction of optical aid. Indeed, looking at the roughness of
Galileo's drawings (by present standards, at any rate), one
might almost suppose that, given the same opportunity, one
probably could have done as well oneself. Yet Galileo's true
skill is nowhere better seen than in a comparison of his work
with that of his contemporaries. Though he made mistakes—
in the case of Saturn, for instance—under the circumstances
he did remarkably well.

Among Galileo's other discoveries, that of the phases of
Venus he recognized to be of fundamental importance in de-
ciding between the rival Copernican and Ptolemaic systems.[8]
Of equal importance was the discovery of four "planets"
around Jupiter which, Galileo proclaimed, had never been
seen "from the very beginning of the world up to our own
time." Their circling about Jupiter was regarded by him as a
beautiful image of the Copernican solar system in miniature.
Kepler wrote to Galileo that he was glad that the four planets—or satellites, as he preferred to call them—had been discovered around a planet rather than around one of the stars.
By doing so, he told Galileo, "you have for the present freed
me from the great fear which gripped me as soon as I had
heard about your book."[9] This fear, he explained, had been of
Galileo's having proved the "dreadful philosophy" of Giordano Bruno, who in 1591 had postulated "planets revolving
around other fixed stars, that is, suns."[10] Bruno had not hesitated to proclaim the universe to be infinite, but even the
usually intrepid Kepler backed away from this, writing that the
doctrine of infinite worlds seemed to imply similar creatures
in each—even infinite Galileos observing new stars in still
other worlds—which he found unappealing, not to say appall-
ing. For his part, he found it preferable to avoid the "march to
the infinite" in the direction of the very large.[11]

Yet Kepler's hopes for a limit in the direction of the large
were, perhaps, partly dashed by Galileo's discovery that the
Milky Way was made up of "a host of other stars, which
escape the unassisted sight, so numerous as to be almost
beyond belief" (the Milky Way was "powdered with stars,"

Milton put it in *Paradise Lost*).[12] The infinite universe Bruno
had envisioned seemed, with this, dramatically confirmed. But
though he recoiled from the dizzying prospect of infinite
worlds, even Kepler had no qualms about assigning inhabitants to the ones already known. "For whose sake," he asked
with respect to the Jovian satellites, "if there are no people to
behold this wonderfully varied display with their own eyes,"
did they exist?[13] For he could not persuade himself that they
were intended only for the delight of the few observers on
Earth who had telescopes.

Tycho Brahe had been equally liberal in assigning inhabitants to the planets, refusing to believe that all these worlds
could be "bare wildernesses existing fruitlessly."[14] In ancient
times the Greek philosopher Democritus had come to the
same conclusion, as had the Roman poet Lucretius, who
wrote in his poem *De Rerum Natura* ("On the Nature of
Things"):

> Now in no way must we think it likely, since towards
> every side is infinite empty space, and seeds in unnum-
> bered numbers in the deep universe fly about in many
> ways driven on in everlasting motion, that this one world
> and sky was brought to birth, but that beyond it all those
> bodies of matter do nought.[15]

He added that there must therefore be "other worlds in other
regions, and diverse races of men and tribes of wild beasts."[16]

Yet though it had thus appeared before, the idea of a plurality of worlds did not really come into its own until the
introduction of the Copernican system. If the Earth was only
one of the planets (as Copernicus had said, and as Kepler and
Galileo had between them done so much to establish), then
did it not follow that the planets must equally be other Earths?
In comparison, the Aristotelian philosophy had lent itself less
readily to extension along these lines, for it had made the
Earth the center of the universe and had considered the starry
heavens a mere shell of flame. Thus, to postulate the existence
of other "worlds" it was necessary to assume nothing less than
the existence of whole other "universes."

The early observers took the Copernican idea of the planets
as other Earths quite literally. The Earth became, indeed, the
schema they most often invoked to interpret their telescopic

images. The number of observers active during the half century or so following Galileo's first observations and the degree of their skill can be fairly well estimated from the series of pre-Huygenian Saturn drawings reproduced in the last chapter (Fig. 2.2), for most of them are represented there.

Among the "strange and marvelous forms" attributed to Saturn, none, perhaps, is stranger than those in the drawings by Francesco Fontana, a Neapolitan lawyer and early telescopic observer. Fontana's drawings of Venus and Mars are just as strange; he described a black "pill" (*pillola*) at the center of their disks and drew fringes of light around them as well. There can be no doubt that such effects were purely optical in origin and were due to the poor quality of Fontana's telescope. Fontana shared the early conviction that the planets were other Earths, and in the case of Venus he believed that he had succeeded in making more concrete the presumed analogy. In 1645 he made out a tiny speck of light above one of the planet's horns and announced on the basis of this that Venus, like the Earth, had a moon. Curiously, the supposed satellite was seen on a number of occasions in the seventeenth and eighteenth centuries by several observers, including Cassini, who was of unquestionable skill. After 1761, however, there were no more reports, and it is now certain that the planet is moonless. In a few cases what had been identified as a satellite was merely a faint star that had happened to creep into the same field with Venus, while Cassini was apparently deceived by optical "ghosts"—internal reflections within the optical system of his telescope.[17]

It is rather surprising that Cassini should have made so basic a mistake. Perhaps the belief that the planets were other Earths had led him to accept his observations less critically than he might otherwise have done. But at any rate, his case proves that even the best observer could be seriously deceived by spurious optical effects. Indeed, progress in planetary astronomy was greatly impeded for some time by the inadequacies of the instruments. Galileo's telescope had employed a simple lens, and all the other early telescopes had made use of the same design; essentially, they were mere "spyglasses." However, as Newton showed in his famous experiment with the prism in 1666, what is perceived as white light is actually composed of the whole spectrum of colors; each color is bent through a slightly different angle by a lens (in other words,

each has a slightly different index of refraction). Red light is
bent less than blue, and thus it comes into focus in a slightly
different plane. When a bright object such as a planet is observed in such a telescope, what one sees is a troublesome mist
of unfocused light around the image—chromatic aberration or
"false color"—and it was this effect that gave rise to the fringes
Fontana erroneously ascribed to Venus and Mars.

By increasing the focal length of a lens, the planes in which
red and blue light come to a focus can be made more nearly to
coincide. Thus the problem of false color could be reduced,
though at the expense of increasing the telescopes' awkward-
ness as they became ever longer. One of Hevelius's had a focal
length of 150 feet (there was no tube; the lens was instead
mounted at the top of a ninety-foot mast). Even this was not
the longest of them—Huygens had one of 210 feet. It is surprising that any useful work could be done at all under such
conditions. As a matter of fact, many of the important discoveries were actually made with telescopes of more modest
length, and at any rate it is clear that the "aerial" telescopes
were more a temporary measure than a final solution. Newton's invention of the reflecting telescope in 1672, which used
a mirror rather than a lens to gather light, showed promise,
but the early reflectors were not particularly effective either,
and until the great William Herschel brought mirror-making
to a higher level of efficiency late in the eighteenth century,
"aerial" refractors and small reflectors tended to be used side
by side—with more or less equally indifferent results.

Speculations about the planets and their inhabitants nevertheless abounded. The *Entretiens sur la Pluralité des Mondes*
(Conversations about the Plurality of Worlds), by the French
writer Bernard de Fontenelle, appeared in 1686, the year of
one of Cassini's observations of the satellite of Venus. "I have
chosen that part of Philosophy which is most likely to excite
Curiosity," Fontenelle wrote in his preface, "for what can more
concern us, than to know how this World which we inhabit,
is made; and whether there be any other worlds like it, which
are also inhabited as this is?"[18] The work became extremely
popular. The *Kosmotheoros*, on the same subject and written
by Christiaan Huygens, was published posthumously at The
Hague in 1698, Huygens having died three years earlier. An

English translation appeared a few years later under the title *The Celestial Worlds Discovered; Or, Conjectures Concerning the Inhabitants, Plants and Productions of the Worlds in the Planets.*[19] Yet practical astronomy had thus far done little to confirm or refute what was discussed in such books. Huygens wrote, "'Tis a very ridiculous opinion that it is . . . impossible a rational soul should dwell in another shape than our own." Yet though he admitted that the inhabitants might look very different, it was impossible even for Huygens to imagine that they would look *altogether* different. After all, they would still need hands to make things and feet to get around. In the same way, it would be difficult for the early observers to imagine that the planets—indistinctly seen in their telescopes—wore other faces than Earth's.

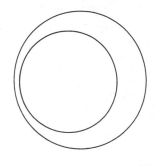

CHAPTER 4

# To Complete
# the Analogy

*With what right, then, does man expect*
*to find artificial or manmade products*
*having even a remote resemblance*
*to those on Earth, on a heavenly body*
*on which even the existence of an*
*atmosphere is disputed?—where the*
*atmosphere, if it exists at all, is so*
*tenuous that wind, precipitation, and*
*the like, are equally impossible?*
                    —W. Beer and J. H. von Mädler,
                                        Der Mond

THE MOON, as the closest of all celestial bodies to the Earth, held out the most immediate promise of a direct discovery of inhabitants, or at least of their handiwork, in another world. William Herschel, writing a century after the *Kosmotheoros* had appeared, commented that "there only seems wanting, in order to complete the analogy, that [the Moon] should be inhabited like the earth."[1]

"Early telescopic observations," the nineteenth-century English astronomer Richard Anthony Proctor wrote, "were conducted with the confident expectation that the Moon would be found to be an inhabited world, and that much would soon be learned of the appearance and manner of the Lunarians." But, he added, "with each increase of telescopic power a new examination was conducted, and it was only when the elder Herschel's great reflector had been applied in vain to the search, that men began to look on the examination as nearly hopeless. Herschel himself . . . was too well acquainted, however, with the real difficulties of the question to share the hopes of the inexperienced."[2]

If Herschel was skeptical that it would prove easy to discover the nature of the Moon's inhabitants by direct observation, he was certainly justified. With more than a hint of sarcasm, Rev. T. W. Webb, author of the well-known *Celestial Objects for Common Telescopes,* compared the expectation of some that such minute details of the surface could be made out with the expectation of the islanders of Teneriffe, "whose simplicity," he said, "led them to imagine that the telescope of Piazzi Smyth would show their favorite goats in our satellite." Webb noted that with a magnification of 1000× the Moon would appear as if it were 240 miles away, "since increase of magnifying power is equivalent to decrease of distance."[3] But he cautioned that "Anyone can judge what could be made of the greatest building on earth at that distance: very small objects, it is true, are discoverable there with the finest instruments, possibly 150 feet broad, or from their shadows one-third as much in height; but their nature remains unknown."[4] Nevertheless, other observers, less cautious and spurred on by the anticipation of a discovery, from time to time reported a false positive.

Germans were the unquestioned leaders in lunar and planetary work in the late eighteenth and early nineteenth centuries, in large part owing to the influence of Johann Hieronymus Schröter, the "father of lunar and planetary astronomy." Born at Erfurt in 1745, he studied law at Göttingen and remained in that profession until the end of his life. However, his official duties were not so taxing as to infringe entirely on his leisure, and he became seriously engaged in astronomical work during the 1770s, about the same time as William Herschel—though unlike Herschel, whose concern was chiefly with the sidereal universe, Schröter's interest was almost entirely in the Moon and the planets. Though he never met Herschel, he did manage to obtain two reflectors built by Herschel, of four-foot and seven-foot focal length, the latter acquired in 1785, some years after he moved to Lilienthal, near Bremen, where he had been appointed chief magistrate. Subsequently he added reflectors of thirteen and twenty-seven feet built by a local telescope maker named Schrader, of which the smaller was proclaimed to be of high quality by the French astronomer Lalande (there are some doubts about the larger). Such instruments allowed reasonably detailed studies of the surfaces of

the Moon and the planets to be undertaken for the first time.
Schröter was almost alone in grasping fully the great opportunity that had opened up, and for more than three decades he
pursued his work with great energy and diligence, the field
being, as Agnes M. Clerke remarked in her *Popular History of
Astronomy*, "then almost untrodden."[5]

A number of Schröter's drawings and descriptions of the
Moon were published as the *Selenotopographische Fragmente* in
1791, and another series appeared in 1802.[6] Schröter firmly
believed in the existence of a lunar atmosphere, having observed, first on February 24, 1792, and then on various occasions over the next nine years, what he took to be traces of a
lunar twilight extending beyond the horns of the crescent
Moon. He also attributed numerous changes he noted in minute surface features to atmospheric or geologic causes. Indeed, the attempt to establish evidence of changes on the lunar
surface was always foremost in his mind, and according to
Beer and Mädler, this

> compromised the value of his painstaking observations
> and drawings. As the outlines of those particular land-
> scapes which seemed to offer nothing pertinent to his
> pursuit of this theme were returned to only superficially,
> he himself could not determine, with certainty, whenever
> a new object appeared, whether it had been visible before.
> Moreover . . . it was only too obvious that he *wanted* to
> notice changes.[7]

Schröter was a clumsy draughtsman—a fact conceded even
by his admirers—though he does not deserve the scorn with
which he has been regarded by some. Of these, Philipp Fauth
is typical. "If his skill as a draughtsman had been at all equal
to his power of observation," Fauth says, "his pictures would
still be valuable." But he went "beyond the problems of his
time" and occupied himself with " 'changes' in the, as yet,
unworked field" rather than laying sound foundations.[8]
Webb's assessment is rather more balanced: "Schröter was a
bad draughtsman, used an inferior measuring apparatus, and
now and then made considerable mistakes; but I have never
closed the simple and candid record of his most zealous
labours with any feeling approaching to contempt."[9] Schröter's opinions about the physical conditions of the Moon led

him to take it for granted that changes took place there and that it was probably inhabited; yet in this respect he was no more unreasonable than Herschel, say, who on one occasion thought he was seeing a lunar volcano in the process of erupting and who held that not only the Moon but also the interior of the Sun was populated with living creatures.[10]

Schröter died in 1816 a broken man, his observatory, including some of his instruments and almost all of his unpublished observations, plundered three years earlier by French troops during the Napoleonic Wars.

The spirit of Schröter's work was taken up by a Bavarian, Franz von Paula Gruithuisen, who had originally been trained in medicine but who later abandoned this for a career in astronomy, becoming in 1826 the director of the Munich Observatory. He undoubtedly had keen eyesight and will always be remembered as among the first to give an account of the meteoritic theory of the origin of the lunar craters.[11] Yet in many ways his views were even more extreme than Schröter's, as suggested by the title of his book, *Entdeckung vieler deutlichen Spüren der Mondbewohner* (Discovery of Many Distinct Traces of the Lunar Inhabitants). He believed that some of the small clefts he had observed on the Moon were "artificial clearings in the forests, answering the purpose of roads,"[12] and he also reported the discovery of a series of "dark gigantic ramparts" (the complex so construed by him was named, fittingly enough, Schröter; it lies on the southern edge of the little dark area known as Sinus Aestum). He said that the ramparts were visible only when near the terminator, and that they were arranged on either side of a central rampart "like the ribs of an alder leaf." Gruithuisen did not believe that the ramparts were currently occupied but instead regarded them as ruins. Later even he found them difficult to make out, but in due time the feature to which he had first called attention was recovered by other observers. Webb, for instance, described it as "a curious specimen of parallelism, but so coarse as to carry upon the face of it its natural origin."[13]

The tendency to exaggerate the regularity of natural features under conditions of difficult observation will prove to be of great importance in what follows. In Gruithuisen's case, his prejudice in favor of an Earthlike Moon no doubt increased the tendency in this direction. But even the cautious Beer and

Mädler noted in the region near the crater Fontenelle (appropriately enough) a "square mountain . . . whose similar walls and rectilinear structure throws the observer into the highest state of astonishment. Despite the very different heights of the sides, their similarity is so great that one can only resist with difficulty the thought that one perceives here a lunar work of art."[14] The appearance of rectilinearity is, however, completely illusory. I have studied the region in question many times myself. In fact, it is only a complex of isolated hills lying in the plain adjacent to the crater.

WHEREAS SCHRÖTER and Gruithuisen undoubtedly were prejudiced in favor of an Earthlike and indeed inhabited Moon, the prejudice of Wilhelm Beer and Johann Heinrich von Mädler was decidedly in the opposite direction. Mädler had been born in 1794 and first became interested in astronomy with the appearance of the Great Comet of 1811. He might have labored in obscurity, however, as a humble seminary instructor had he not met Beer. Three years Mädler's junior (though his name always appears first when their joint work is mentioned), Beer was a well-to-do Berlin banker of a distinguished Jewish family. Beer had an interest in astronomy, and Mädler began to give him lessons. Being a man of considerable means, Beer set up his own observatory by the famous Berlin Tiergarten, equipped it with an excellent 3¾-inch refractor made by the great optician Fraunhofer, and with Mädler at once set to work making observations for their great map of the Moon's surface, generally using a magnification of 300× when drawing. The map appeared in four parts between 1834 and 1836, and the year after its completion they published a huge book, Der Mond, in which they attempted to interpret the observations.[15]

It has always been recognized that Mädler did the greater share of the work. After Mädler left Berlin in 1839 to take up a position as director of the Dorpat Observatory in Estonia, Beer seems, indeed, to have given up astronomical work altogether. All the same, it is not fair to underestimate Beer's contribution: Mädler's later work also never attained the same high level that it had during the years of their collaboration.

Der Mond was acknowledged at once as a scientific classic. The great Alexander von Humboldt spoke highly of it; Mädler received the honorific title of Royal Professor of Astronomy from the king of Prussia and was invited by the director of the Berlin Observatory, Johann Franz Encke, to observe with the 9.6-inch refractor there. Mädler's later appointment to the directorship of Dorpat, one of the leading astronomical establishments of Europe, was also chiefly in recognition of his lunar work.

Beer and Mädler were clearly disgusted with the extravagant claims of lunar changes that had been made on the most questionable grounds. Observers such as Schröter and Gruithuisen, fascinated by the remarkable transfigurations of the features rendered by ever-changing conditions of light and shade, had found it only too difficult to escape the force of such overwhelming impressions. As Webb beautifully described the lunar scene:

> Objects recently well recognised under the effect of light and shade will become confused by a novel effect of local illumination, and the eye will wander over a wilderness of streaks and specks of light, and spots and clouds of darkness. . . . Unknown configurations will stand boldly out defying all scrutiny, and keeping their post immovably till the decreasing angle of illumination warns them to withdraw. Nothing can be more perplexing than this optical metamorphosis, so complete in parts as utterly to efface well-defined objects; so capricious as in some instances to obliterate one, and leave unaffected the other, of two similar and adjacent forms. Gruithuisen, carrying out an idea of Schröter's, referred some of these changes to the progress of vegetation which, if existing, will naturally, in default of a change of seasons, run its whole course in a single lunation.[16]

The growth and decay of lunar vegetation was supposed to take place over the course of a single lunar day and night, since the Moon, having only a slight inclination of its axis ($1°32'$), has nothing else comparable to the terrestrial seasons. But whatever else they may have been, such conjectures were premature. Beer and Mädler soberly rejected most of the supposed variations as due entirely to changing illumination of the surface; in this they were certainly prudent.

Not only were the lunar surface features apparently static

and unchanging, Beer and Mädler concluded, but their observations seemed to negate change of an atmospheric nature as well. To Beer and Mädler the lunar limb always appeared perfectly sharp, and they sought in vain for any extensions of the horns such as those Schröter had alleged. Moreover, the lunar shadows were always perfectly chiseled at their boundaries, in contrast to the softer shadows of our own world, where the sunlight is diffused by an atmosphere. Though the rims of craters, catching the sunlight, might shine dazzlingly bright, their floors were always profoundly black. By taking special precautions, the skilled observer Rev. William Rutter Dawes, it is true, was later able to detect some ghostly outlines in the floors of certain craters, but he rightly recognized that these outlines were merely faint glows due to the reflection of light from the nearby cliffs. Of lunar atmosphere there was, clearly, little or none, and in that case there could be no liquid water either. Consequently the "seas" could not be seas. To quote Beer and Mädler's own words,

> As the assumption of a lunar atmosphere stands or falls, so do all assumptions about clouds and smokelike exhalations, nebulous veils, precipitations, and the like—along with any water cycle, and water itself. The conclusion follows that there must be an absolute difference in every respect. Specifically, the economy of organic nature must be as different as the surfaces. The Moon is no copy of the Earth, and even less is it a colony of the Earth. Planetary and lunar forms of life cannot be compared, and a further discussion of the question, "whether the Moon is inhabited by *men*," will very likely seem superfluous.[17]

It is often thought that it was the landing of men on the Moon that finally destroyed its romance; actually, Beer and Mädler's observations did so long before. Thinking, no doubt, of Beer and Mädler, the poet Tennyson wrote: "Dead the new astronomy calls her."[18] The scientific situation after the appearance of Beer and Mädler's book in 1837 was well summed up by Richard A. Proctor:

> The examination of mere peculiarities of physical condition is, after all, but barren labor, if it leads to no discovery of physical condition. The principal charm of astronomy, as indeed of all observational science, lies in

the study of change,—of progress, development, and decay, and specifically of systematic variation taking place in regularly recurring cycles. . . . In this relation the Moon has been a most disappointing object of astronomical observations. For two centuries and a half, her face has been scanned with the greatest possible scrutiny; her features have been portrayed in elaborate maps; many an astronomer has given a large portion of his life to the work of examining craters, plains, mountains, and valleys for signs of change; but hitherto no certain evidence—or rather no evidence save of the most doubtful character—has been offered that the Moon is other than "a dead and useless waste of extinct volcanoes."[19]

It is certainly no coincidence that when a resurgence of interest in the Moon finally came, it was only after a reasonably creditable case for what Proctor calls "physical condition" had been made. In 1866 Julius Schmidt, a German astronomer and director of the Athens Observatory, startled the world with his announcement that Linné, a small crater on the dark plain Mare Serenitatis had apparently changed.[20] Wilhelm Lohrmann, who had commenced work on a map of the Moon before Beer and Mädler but who had had to give it up because of failing eyesight, described Linné as a *deep* crater more than six miles across; Beer and Mädler had concurred. Schmidt's own observations dating back to the period 1840–43 were consistent with this description. However, in 1866 he noticed, with the 6-inch refractor at Athens, that instead of a deep crater what appeared was rather a brilliant whitish cloud. There was no trace of any crater per se. The following year Father Pierre Angelo Secchi at Rome detected a much smaller crater some two miles across at the center of the whitish cloud. The earlier crater, it was assumed, had been largely obliterated by a lava flow, leaving this smaller crater in its place.

Schmidt had been almost alone in being undaunted by Beer and Mädler's achievement. Indeed, his lunar observations had begun in earnest the very year that their great book appeared (it may not be amiss to mention here that his lifelong interest in the Moon began when he received, while yet a boy, a copy of Schröter's *Selenotopographische Fragmente*). Following Schmidt's announcement, a controversy over the alleged change in Linné raged for many years. It has now been settled once and for all: spacecraft photographs clearly show it to be an ordinary, albeit a relatively fresh, impact crater.[21] Ironically, the often-maligned Schröter, in a drawing he made on November 5, 1788, had shown it essentially as it looks today. Schmidt was therefore mistaken in this case. But his mistake only testifies to the extreme difficulty of Schröter's project of detecting genuine changes *visually*, given the differing conditions of illumination and shading of the surface features, not to mention the equally important differences in observers' instruments, eyes, and artistic mannerisms. To sort out any genuine alterations from the merely supposititious was—we can say in retrospect—well-nigh impossible.

Though the discussion about minor changes—an occasional obscuration, or a volcanic outgassing every now and then—would continue, and indeed goes on to the present day, it was impossible to return to the total naivete about lunar conditions that had prevailed in the days before Beer and Mädler, when Kepler had confidently proclaimed that the dark regions were seas and the light areas continents, or when Herschel had called attention to the striking analogy between the Earth and the Moon, which only needed the existence of lunar inhabitants to be complete. Now Webb wrote instead of the "want of terrestrial analogies," and added that

> it may be reasonably supposed that Venus or Mars, at the like distance, might be far more intelligible. We should certainly not find them mere transcripts of our own planet, for the conditions of existence as to temperature and air, if not in other respects, are very different, and, as Schröter often remarks, variety of detail in unity of design is characteristic of creation; but we might have a fair chance of understanding what we saw. It is quite otherwise with the Moon. It is, in Beer and Mädler's words, no copy of the Earth; the absence of seas, rivers, atmosphere, vapours, and seasons bespeaks the absence of "the busy haunts of men"; indeed of all terrestrial vitality.[22]

"No copy of the Earth"—from Kepler's "The spots are seas, the light areas are lands" to this took a little more than two centuries. One sees the usual psychological progression here.

Faced with an unfamiliar object, one at first tries to classify it within some framework with which one is already familiar (in this case that of the Earth, with its oceans and land). Having thus made a start, one proceeds to make adjustments in order to arrive at a better fit. "It is only by comparing that we can judge," wrote George Louis Buffon in his *Histoire Naturelle de L'Homme* of 1749, "since our knowledge rests entirely on the relation that things have with others that are similar or different."[23]

The general process, which Sir Ernst Gombrich has described as making up "the cycles of schema and correction," one sees in the artist's first sketch, which captures the form only loosely, or in children's drawings, which are only concerned with getting the framework right and which ignore the details.[24] The child's singular attention to what is most essential in characterizing a figure is what gives his or her drawings their abstract quality and makes them humorous to adults, who are struck by the fact that "the man has no ears." Soon enough the child will learn to remedy that. It is, evidently, the same process that has been at work in the history of art since the beginning. "Painting begins its career," wrote Percival Lowell in *The Soul of the Far East,* "in the humble capacity of copyist, a pretty poor copyist at that. At first so slight was its skill that the rudest symbols sufficed. 'This is a man' was conventionally implied by a few scratches bearing a very distant relationship to the real thing. Gradually, owing to human vanity and growing taste, pictures improved. Combinations were tried, a bit from one place with a piece from another."[25]

If, as has been suggested, the unfamiliar is to be grasped only in terms of what is already familiar, a certain provincialism, not to say egotism, is inevitable in the initial schema. Assuredly there is within us the desire and the tendency (no doubt infantile) to see everywhere in the nonhuman world the "human face divine." Hence we speak of nature as "Mother" and of the sky god as "Father." The child, and to a lesser extent the "primitive," believes everything to be sentient—seeing faces in stones, trees, clouds; carving faces when nature does not do well enough. Thus is born sculpture and religion (God made man in his own image, and by inference God must therefore be made in man's). And no doubt poetry has a similar genesis: the poet (in this respect infantile or primitive) feels compelled to let his own sentience radiate forth until it vivifies all, until the Moon takes on a countenance and the stars "glare."

It is perhaps but an extension of this desire and tendency to see everywhere in nature the human face that when astronomy advanced to the recognition that the wandering lights in the sky were worlds it should seek to find in every world the image of the face of the Earth, with oceans and lands.

Almost without exception the early observers saw "the world in the Moone" that Bishop Wilkins had written about in 1638. *Almost* without exception, for it is worth noting that Christiaan Huygens, while supporting the general notion of the plurality of worlds in his *Kosmotheoros,* had bypassed the Moon because, he said, "it has no seas, no rivers, nor clouds, nor air and water."[26] But it had taken Beer and Mädler's careful work to establish once and for all that the Moon was "no copy of the Earth" and that henceforth there would be two worlds in terms of which one could look at planets still farther afield. A first glimmering of the Earth's uniqueness had thus come about only with the recognition of the Moon's otherness. The repertoire of schemata had grown from one to two. (In our own day, of course, we have seen the repertoire grow still more, until at last we have become true students of what might be called the comparative literature of worlds).

By the mid-nineteenth century, attention was turning to more remote Venus and Mars. Though Webb cautioned that neither of these could be expected to be exact "transcripts" of the Earth, what appeared in this first stage of further discovery certainly suggested a nearer resemblance to the Earth than to the Moon. Of Venus, Webb said: "Could we but see [its] features more readily, what an interesting object would this lovely planet become, especially as in point of size it is the only companion to the Earth in the whole system."[27] Despite the less perfect resemblance in size, Mars seemed to be still more like the Earth, prompting William Herschel to write in 1784: "The analogy between Mars and the earth is perhaps by far the greatest in the whole solar system."[28] Father Secchi, friend and mentor to Schiaparelli, was later to agree, using words reminiscent of those Kepler had spoken with respect to the Moon two centuries before: "Everything is variegated like a map of the Earth."[29]

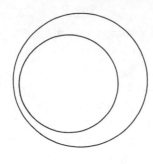

## CHAPTER 5

# Like a Distant
# View of Earth

*The distant view of the Earth, indeed,
might be much of this nature; its
features at one time distinct, at another
confused or distorted by clouds.*
*—Rev. Thomas William Webb,*
Celestial Objects for Common
Telescopes

AFTER THE MOON, Venus approaches closest to the Earth, and the splendor with which it dazzles the naked eye as the evening or morning "star" could not fail to attract the attention of the early telescopic observers. Perhaps some of them felt something of that sentiment later expressed by the poet Tennyson, with its hint that the loveliness might, after all, be illusory:

Venus . . . smiling downward at this earthlier earth of ours,
Closer on the Sun, perhaps a world of never fading flowers.

Hesper, whom the poet call'd the Bringer home of all good
    things.
All good things may move in Hesper, perfect peoples,
    perfect kings.

Hesper—Venus—were we native to that splendour or in
    Mars,
We should see the Globe we groan in, fairest of their
    evening stars.

Could we dream of wars and carnage, craft and madness,
    lust and spite,
Roaring London, raving Paris, in that point of peaceful
    light?

Might we not in glancing heavenward on a star so silver-fair,
Yearn, and clasp the hands and murmur, "Would to God
   that we were there"?[1]

The planet's very brilliance makes it a particularly unsuitable
object for telescopes of dubious optical quality. Cassini, at
Bologna, sketched a few spots in 1666–67, but after leaving
his native Italy for Paris on being appointed director of the
observatory there, he never recovered them. From the Italian
observations he had arrived at an Earthlike rotation period of
about 23 hours.[2] When Francesco Bianchini later observed the
planet with a 2½-inch refractor of 66-foot focal length at
Rome, he found a rotation period of 24 days, 8 hours. Bianchini sketched various dusky patches on the planet, which
he thought definite and permanent enough to allow him to
draw a map; it showed, not surprisingly, "continents" and
"oceans."[3]

Schröter took up the planet in 1779 but failed to see anything at all on the disk for nine years. Finally, on February 28,
1788, he detected a "filmy streak." Over the next two years he
described the southern horn as blunted, and he noted what he
believed to be the summit of a mountain just over into the
darkened part, the height of which he estimated at 27 miles.[4]
Herschel, who had had difficulty making out anything on
Venus, wrote sharply: "As to the mountains on Venus, I may
venture to say that no eye, which is not considerably better
than mine, or assisted by much better instruments, will ever
get a sight of them."[5] Schröter, though hurt by Herschel's
criticism—it was, he said, such as he "could not reconcile
. . . with the friendly sentiments the author [had] always hitherto expressed towards me"[6]—nevertheless continued to observe the planet, and he discovered still other irregularities of
the terminator. In retrospect we can say that he was right
about the irregularities; they have been seen on occasion
since.[7] But he was wrong about their interpretation. Clouds as
well as mountains can produce such effects (ironically, Schröter's own observations did much to prove the existence of a
dense atmosphere around Venus).[8]

Schröter essentially agreed with Cassini's rotation period
for Venus, publishing several values all around 24 hours. His

final result, which appeared in 1811, was 23 hours, 21 minutes, 7.977 seconds.

Schröter's irregularities of the terminator of Venus were probably authentic. But what is one to say of the observation of P. La Hire in 1700 that "the irregularities of the terminator of this planet are very much more considerable than those of the Moon"?[9] Or of the valley surrounded by mountains and described as "resembling a lunar crater," which Father De Vico and his assistants at Rome saw march steadily, night after night, toward the planet's terminator in March and April 1841?[10] Such observations suggest an unconscious tendency to construe Venus in terms of the Moon—a tendency that the analogy of the phases may partly explain. On the other hand, the rotation period ascribed to the planet by every observer except Bianchini bespeaks an equally strong tendency to see the planet as another Earth. In particular the period found by De Vico and his assistants—six observers in all—of 23 hours, 21 minutes, 22 seconds (only fourteen seconds longer than Schröter's) is certainly remarkable, given the fact that the rotation of Venus was later to be regarded as one of the most uncertain desiderata of planetary astronomy. It illustrates once again that process which appears so often in the work of the early observers: their tendency to see what they expected to see.

As great as had been the promise it had held out for showing similarities to the Earth because of its size, Venus did not, unfortunately, prove a very satisfactory object of close telescopic scrutiny. Its markings were regarded, by most observers at least, as remarkably ill-defined, which was, to such observers as Herschel and Schröter, suggestive of a dense atmosphere. Webb was one of few nineteenth-century observers to take Bianchini's oceans and continents seriously. As we have seen, he wistfully noted: "Could we but see these features more readily, what an interesting object would this lovely planet become." But in fact they defied exact delineation.

Like the Moon after Beer and Mädler's careful study, the planet came to be relatively ignored, though for precisely opposite reasons. Whereas in the case of the Moon it seemed that our knowledge of it was complete, our knowledge of Venus was rather hopelessly incomplete. Nor was it possible to foresee how this was likely to change. If the planet was covered

pole to pole with clouds, they offered no prospect of parting
to reveal for even a moment anything of the underlying surface. Like the beauty who, while splendid to look upon, is dull *Like a*
to know, the planet so dazzling to the naked eye offered rela- *Distant View*
tively little inducement to further acquaintance, being too *of Earth*
often devoid of interest except for its characteristic phases. The
markings of its face, mere nebulous patches of uncertain outline and duration, could not even be confidently affirmed as
real, it being reasonable to say of them, borrowing a phrase of
Virgil, "one sees, or thinks he has seen."[11] To the visual observer the planet was a cipher.

MARS, the next nearest neighbor in space after the Moon
and Venus, was completely different. True, it still lay outside
the pale of clear visibility; yet it was neither so far away nor so
entirely veiled as to be utterly devoid of definite details. What
was seen was, while not altogether plain, suggestive and perhaps even familiar. But the view was still only suggestive of
possibilities. Though power tends to corrupt, as Lord Acton
remarked, there was not an astronomer alive who did not wish
desperately for an increase of power to bring the planet closer,
to make its ambiguous outlines stand forth sharper and better
defined.

The tantalizing ambiguity of this planet was, of course, in
the first place the result of the mere accident that size and
distance together make it just large enough as seen from Earth
to show some details and just small enough not to show those
details too clearly. While its diameter is only twice that of the
Moon, Mars can never approach nearer than 140 times the
Moon's distance, so at best its disk appears only about as large
as a medium-sized lunar crater. Moreover, owing to the situation of its orbit relative to the Earth, it can be studied to best
advantage only at intervals of fifteen or seventeen years.

As the Earth and Mars travel in their respective orbits about
the Sun, at more or less regular intervals they line up on one
side and thus minimize the distance between them. The time
between these alignments—called *oppositions* because Mars
then appears opposite the Sun in the sky as seen from Earth—
is on average about 2 years, 2 months. If the orbits of Earth
and Mars were exactly circular, the distance between the two
worlds would be the same at each opposition; however, since

they are in fact elliptical, as Kepler showed, some oppositions will be more favorable than others. The orbit of Mars differs from a circular shape more than that of the Earth; therefore, it is Mars's position in its orbit which is the chief factor in determining the opposition distance between the two. If Mars is near perihelion (the point of its orbit at which it is nearest the Sun), the distance may be as little as 34.6 million miles (55.7 million kilometers); if it is near aphelion (the point of its orbit farthest from the Sun), the distance may be as much as 62.9 million miles (101.3 million kilometers). The favorable oppositions always occur in either August or September, because the Earth passes the perihelion of the Martian orbit in late August each year; the least favorable occur in February or March, the Earth making its passage by the aphelion of the Martian orbit in late February.

The difference between the most favorable and the least favorable oppositions is considerable. At the August or September oppositions, the planet's area in the telescope is three times as large as at the February or March ones. The August and September oppositions, which occur at fifteen-year intervals, have therefore tended to be watershed years of Martian discovery, though the advantage of distance has to some extent been offset for the northern hemisphere, where most of the great observatories are situated, by the fact that Mars then has a far southerly declination and must be observed low in the sky.

IN SPITE OF the difficulty in making observations with the clumsy "aerial" telescopes of the seventeenth century, even in those early days observers had already glimpsed a few details on the Martian surface. Christiaan Huygens, on November 28, 1659, was responsible for producing "the first drawing of the planet worthy of the name," as Percival Lowell later described it.[12] The drawing (Fig. 5.1) shows in clearly recognizable form the hourglass outline of a darkish region against the predominantly ochre background of the disk—the Syrtis Major, as it came to be known on later maps. It seemed reasonable to suppose that this might be a great sea in the midst of lighter lands—and indeed, as had been the case with the Moon, by analogy to the Earth the light regions were assumed to be lands and the dark areas seas.

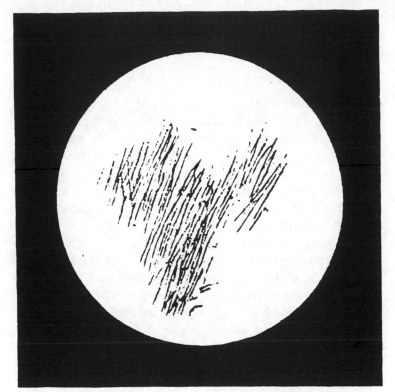

Figure 5.1
Christiaan Huygens's view of
Mars, which Percival Lowell called
"the first drawing of Mars worthy
of the name." The dark area
shown was long referred to as the
"Hourglass Sea" but is now known
as Syrtis Major.

A few years later, in 1666, Cassini made out whitish regions at the planet's poles whose brilliance could only be due, it seemed, to the gleaming of ice. This made still more convincing the resemblance that first impressions suggested between Mars and the Earth.

An important series of observations begun by William Herschel in 1777 established further that the planet's obliquity, or the angle by which the equator is tilted to the plane of its orbit, is at 25°, just about the same as Earth's. This fact was added to the agreement, known since Huygens's time, between the periods of rotation of the two worlds. Thus Mars seemed not only to have lands and seas, ice caps at its poles, and Earthlike periods of day and night, but also to experience Earthlike seasons, leading Herschel to conclude that "the inhabitants probably enjoy a situation similar to our own."[13]

This, then, was the tantalizing figure that Mars had already begun to cut when Beer and Mädler turned their 3³/₄-inch refractor toward the planet at the unusually favorable opposition of 1830.[14] Their telescope, though small, was of outstanding quality. It was one of the achromatic refractors that were revolutionizing astronomy, and planetary astronomy in particular, during the nineteenth century. Because of the problems with "false color" that had plagued the early refractors and that had led to the "aerial" telescopes, Herschel had turned wholeheartedly to reflectors. An Englishman, John Dolland, was at the same time pursuing a solution along different lines. He found in 1759 that by making a compound lens, of which one component was of crown glass and the other of flint glass, the more serious problems of "false color" could be eliminated. Unfortunately, Dolland's achromats were all fairly small, because at the time it was impossible to cast striation-free flint glass disks more than about 4 inches in diameter. By early in the next century, improvements in glass-making techniques had made it possible to go beyond this, and the achromat, in the hands of the skilled Bavarian optician Joseph von Fraunhofer, was brought to near perfection. His masterpiece was the lens of the 9¹/₂-inch refractor at Dorpat, in Estonia, but Beer and Mädler's 3³/₄-inch was of the same high quality. There was still some chromatic aberration, admittedly, but it was now much less of a nuisance than in the early days.

Beer and Mädler drew the outlines of the Martian "lands" and "seas" with a surer hand than had Herschel or any other of their predecessors, producing drawings that began to show them in consistently recognizable form (Fig. 5.2). (Schröter, by comparison, had failed to convince himself that the same features were always visible, and he concluded that the planet was surrounded by a "shell of clouds.")[15] The German astronomers continued their observations at the subsequent oppositions of 1832, 1834, and 1837, and by 1837 they regarded the permanence of the various spots as established. In light of the fact that they could not be influenced—as later observers arguably were—by seeing better maps and knowing what to expect beforehand, they ought to be forgiven a certain amount of distortion, particularly as their telescope was small and the Martian features viewed in it showed up only indistinctly. Indeed, as they themselves noted, "usually some time had to elapse before the indefinite vague mass at first seen resolved itself into clearly distinguishable forms."[16]

Curiously, after obtaining further observations in 1839 and 1841 with the 9½-inch refractor at Dorpat, Mädler began to doubt the permanence of the spots and returned to the view of Schröter that the markings were all clouds. To subsequent observers, for whom the evidence of their permanence had become overwhelming, the true explanation seemed obvious. As Webb wrote, "The distant view of the Earth, indeed, might be much of this nature; its features at one time distinct, at another confused or distorted by clouds."[17] The observations therefore seemed to point still more compellingly to an Earth-like Mars. This was no Moon, stoic and unchanging, but a world where atmospheric mists and clouds were apparently capable of rendering the familiar outlines unrecognizable for periods, just as the clouds in the Earth's atmosphere might be presumed to do for an observer at so great a distance.

In 1837, Beer and Mädler had noticed that the north pole of Mars "appeared surrounded by a conspicuous dark zone," which they thought might possibly be a marsh at the edge of melting snow. It was seen again, though not so clearly, by Mädler in 1839 but not in 1841. At these times the north pole was tilted toward the Earth, so it could not be determined whether the south pole exhibited the same appearance; in 1856, Webb found that it did.[18]

Figure 5.2
Observations of Mars by the
German astronomers Wilhelm
Beer and Johann Heinrich von
Mädler, 1830 – 32, using Beer's
3 ³⁄₄-inch Fraunhofer refractor.

Beer and Mädler were certainly under no illusion that what they had accomplished was other than a tentative beginning, as testified by the fact that on their maps of the planet they did not even bother to name the various features; they merely assigned them letters. This provisional nomenclature was not particularly convenient. A different scheme was proposed in 1867 by the English astronomer Richard A. Proctor,[19] in which features were named after astronomers who were in many cases still living and who were predominantly English (Fig. 5.3)—so that it could be said of Proctor's Mars as it was of the Earth in those days: "The Sun never sets on the British Empire." Of the forty-odd names on Proctor's map, fully half were those of his countrymen; moreover, some were used more than once. That of the English astronomer William Rutter Dawes appeared no less than five times: in Dawes Ocean, Dawes Continent, Dawes Sea, Dawes Strait, and Dawes Forked Bay.

The reason for Dawes's prominence on Proctor's map was that it had been largely on the basis of his exceptional series of Martian observations, made with an 8¼-inch refractor at the fine opposition of 1864, that Proctor had drawn it, though certainly Proctor's crude rendition does not do justice to Dawes. For besides having one of the keenest eyes for the discernment of planetary detail (he was known as "the eagle-eyed"), Dawes was also a skillful draftsman. His drawings (Fig. 5.4) are superior not only to those of Beer and Mädler before him but also, in their representation of subtle differences of shade in Martian features, to the hard and sharp representations of such celebrated later observers as Schiaparelli and Lowell.

The son of a mathematics teacher who had served as an astronomer on an expedition to Botany Bay in Australia in 1787, Dawes was born in London in 1799.[20] His father intended him for a career in the Anglican church, but the youth was unable to reconcile himself to all of its tenets and prepared instead for the medical profession, eventually setting up a practice in Liverpool. While there, however, he fell in with an Independent preacher, Thomas Raffles, who persuaded him to take charge of a small Independent congregation in the village of Ormskirk, north of Liverpool, so he became a clergyman after all.

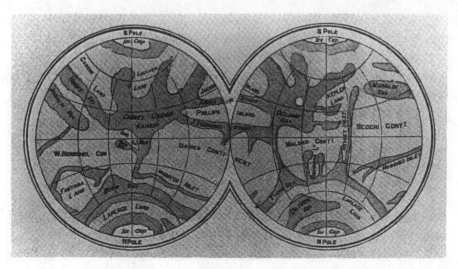

*Figure 5.3*
*Map of Mars by Richard Proctor.*
*This is an 1871 version of Proctor's*
*original 1867 map. Some of the*
*names that gave rise to an*
*overrepresentation of British*
*astronomers have been dropped to*
*make the map more palatable to*
*Continental astronomers. The*
*name of Dawes, for example,*
*appears only twice rather than five*
*times, as in the 1867 version. The*
*name Malder is a misprint of*
*Madler.*

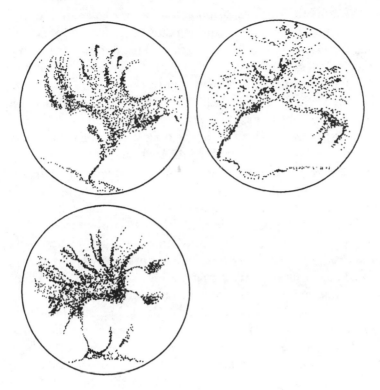

*Figure 5.4*
*Drawings of Mars by Rev.*
*William Rutter Dawes, 1864–65.*
*These rank among the most*
*accurate representations of the*
*Martian surface produced during*
*the nineteenth century. Note the*
*streaky features in the drawings,*
*which prefigure the Schiaparellian*
*canals.*

Meanwhile, Dawes's astronomical interest had surfaced, and he began to make his own observations, especially of double stars, a line in which he was to excel. His first instrument was only a 1.6-inch refractor, but what he lacked in equipment he made up for in zeal. In the last year of his life he reminisced to his friend Sir John Herschel: "I worked away on almost every fine night, when uncertain health would permit, and found and distinctly made out . . . [the double stars] Castor, Rigel, $\epsilon^1$ and $\epsilon^2$ Lyrae, $\sigma$ Orionis, $\zeta$ Aquarii, and many others, of which I made correct diagrams in a book now lying before me. . . . The difficulty was often to get to bed in summer before the sun extinguished the sight of the game."[21]

Dawes mentions his uncertain health, and indeed it was never good, as he suffered from asthma and headaches. After the death of his first wife in 1839 it broke down, and he was forced to give up his congregation. He then worked for a time as an assistant at the private observatory of a wealthy businessman, George Bishop, at St. John's Wood, London. Finally, in the 1840s, he achieved by his remarriage to a well-to-do solicitor's widow the financial independence he needed to establish himself at a country estate in Kent, where he equipped himself with a fine $6^1/_3$-inch refractor by Merz (Fraunhofer's successor). It was this telescope that he used in 1850 to discover the inner or "crêpe" ring of Saturn (independently of W. C. and G. P. Bond, who had found it two weeks earlier at Harvard with the 15-inch Merz refractor there). Incidentally, Dawes's health had meanwhile taken such a turn for the worse that he had seriously considered giving up astronomy just before making this, his most celebrated discovery.

Though in later years Dawes turned to the planets with signal success, double stars remained his abiding interest. In the course of his observations of them, made over thirty years with a variety of instruments, he became intrigued by the question of how a telescope's ability to separate close double stars—its resolving power—depends on its aperture. The question of resolving power is quite as pertinent to the observation of planets, for clearly what a telescope will show on a planet's surface is no less critically dependent on the aperture. To discuss the matter properly requires a brief digression into optics.

Consider the light from a distant star, which for all practical purposes can be thought of as coming from a point source.

The stars are so far away that though we know that they are actually suns in their own right and that some of them are of truly enormous size, their disks are much too small to be seen directly in even the largest telescopes. Nevertheless, the stars *do* show small apparent disks when viewed through a telescope and paradoxically, the smaller the telescope, the larger the apparent disk.

The reason for this has to do with the wave theory of light. If a stone is cast into a pool, a circular wave is produced which emanates outward from the point at which the stone entered the water. The wave's origin is called the source, and the front along which it moves outward is called the wave front. In 1678, Huygens proposed that light itself has the character of a wave. He introduced a principle by means of which, provided the initial position of a wave front is given, its position could be found for any subsequent time. This principle could be used to derive some of the basic laws of optics, such as those of reflection and refraction.[22] One considers each point along the initial wave front as itself a new source of disturbance, which produces a wavelet that emanates outward from it; the wave front that results is found by drawing the tangent to the envelope of wavelets.

There were certain problems with Huygens's theory. To begin with, taking each point on the wave front as a secondary source, as prescribed by Huygens's principle, the wavelets would be expected to travel backward as well as forward. However, only forward propagation occurs. The theory also implies that light should be able to bend around corners to some extent, as sound waves do, but this had not been observed. Light instead appeared always to travel in perfectly straight lines (thus in geometrical optics one spoke of *rays* of light).

It was later shown, however, that the bending of light around corners does occur; it had simply not been observed by the early workers, owing to the short wavelength of light. Light waves were also observed to interfere with one another, as in Thomas Young's famous two-pinhole experiment of about 1800. Where the crest of one wave coincided with the crest of another, they reinforced one another and produced a bright band; where the crest of one coincided with the trough of another, they canceled out, producing (strange as it seems) darkness.

The bending of light around corners is called diffraction. If light traveled in perfectly straight lines, a point source of light could be imaged as a point. But of course if light is a wave, it will not travel in perfectly straight lines. This means that even if the source is a point, the image will have a certain finite size because of the bending of light waves around the edge of the lens. What one looks at when one turns a telescope toward a star (which for all practical purposes may be regarded as a point source) is therefore actually the diffraction pattern produced by the star. This consists, in the case of the image formed by a circular aperture, of a bright central disk containing 84 percent of the star's light, surrounded by a series of narrow rings produced by interference.[23] The problem was first worked out mathematically by George (later Sir George) Airy, the English astronomer best remembered today, alas, as the villain in the story of the discovery of Neptune.[24] Airy found that the radius for the central disk (known as the *Airy disk*) formed by a circular aperture is given by the formula,

$$\theta_r = 1.22 \frac{\lambda}{a}$$

where $\lambda$ is the wavelength of light and $a$ the diameter of the aperture. If compatible units are used in the numerator and denominator, the result will be expressed in radians, where 1 radian = 206,265" of arc.

Dawes had already found out from his own experiments with various apertures that the size of the central disk in the diffraction pattern is inversely proportional to the aperture of the telescope—the gist of Airy's formula above. He also noted that if the diffraction patterns of two stars as produced in a given telescope overlap too much, it becomes impossible to distinguish them; they remain unresolved. He found that a 1-inch aperture would just separate a double star composed of two stars of the sixth magnitude when their central distance was 4.56 seconds of arc. From this he defined what is now known as Dawes' limit: the separating power of any given aperture, $a$, will be expressed by the fraction 4.56"/$a$, where the aperture is expressed in inches (conditions of reasonably steady atmosphere being assumed).[25] It should be emphasized that this result, announced by Dawes in 1865, is a completely empirical one, established for small refractors and for his keen eye; other observers might arrive at a somewhat different limit.

Though first defined during the course of work with double stars, the diffraction limit is just as important to planetary observation. This may be grasped by analogy to the photographs printed in newspapers. On examination with a magnifying glass, such photographs will be seen to be made up of many small dots, or pixels; the smaller the pixels, the smoother the photogragh appears. Diffraction may be thought of as determining the pixel size of the image of the planet in the telescope; if the pixels are too coarse, applying higher power is useless because all one sees are the individual pixels magnified, not details smaller than the pixel size.

Since a bright point on a planetary surface will blow up into a disk $2\theta r$ in diameter as seen in any telescope, two such points, if the aperture is insufficiently large, will not be separated but instead will blur into a single large bright spot. A dark spot may be thought of as bounded by a series of bright points all around its circumference, each of which will form an Airy disk; the dark spot will therefore be encroached upon on all sides, and it will appear correspondingly shrunken. Similarly, a dark stripe will appear narrowed, owing to the same sort of "bleeding over" of bordering bright areas. In summary, the results of our analysis may be expressed as follows:

1. Bright points on a planetary surface or bright bands should appear broader when imaged in a small telescope than in a large one.
2. Dark spots and dark bands should appear narrower.

BUT FOR THE wave character of light, the image formed by any telescope would be equally capable of withstanding high magnification and of showing minute planetary surface details. However, in the real world, light is not so obliging; its wave character, and the diffraction that is a consequence of it, limits the magnifying power that can be fruitfully employed—and the smaller the aperture, the more it limits it. This explains the desire of planetary observers to apply larger and larger apertures to their study. Unlike observers of faint stars and nebulas, they did not desire to obtain *more* light but rather to push back the limitations to their seeing imposed by diffraction.

It might appear that the telescopic study of a planet must be simply a matter of bringing larger and larger apertures to bear on it, that what is undetermined in a small telescope need only be examined in a larger one to have its true nature defined. To some extent this is true, but there are other waves as pertinent as those of light to planetary observation. These are atmospheric waves; and just as light waves impose stricter limits on smaller telescopes, atmospheric waves were apparently more handicapping for larger ones. The observations of the early workers, with their small apertures, were effectively limited by diffraction alone, but when very large instruments began to be built, it became increasingly obvious that atmospheric conditions often played as critical a role as diffraction in planetary seeing and that the advantage of superior apertures was not as great as might have been anticipated. A rather heated controversy arose toward the end of the nineteenth century between users of large telescopes and users of relatively small ones concerning which kind of instrument was actually most suitable for planetary work.

At the moment, all this still lay in the future. With Dawes's superb drawings of Mars in the 1860s, the main outlines of the Martian "seas" and "continents" (as they were then widely believed to be) had been depicted about as well as humanly possible by Dawes's uncanny skill; the coarser features of the Martian surface, so imperfectly revealed to Huygens and the other early observers, were clear enough. But on those same drawings on which the broader features of the Martian disk were so faithfully represented, there were also tentative indications of a class of finer details: the elongated outlines of the dark "seas" seemed to prolong themselves, here and there, from narrow baylike extensions into thin, wispy streaks, which gradually disappeared into the ochre parts of the planet (hitherto regarded as merely a "ground" against which the "figures" of the dark areas could stand out, to use the terminology of the Gestalt psychologists). This was significant, for it showed that, as completely as the broader features had now been mapped, there was yet something more awaiting discovery: a fine print that was just starting to be made out and in relation to which humankind now stood, in the 1860s, in the same relation as Huygens had stood to the big print two centuries before. Perhaps when Mars next swung around to one of its exceptionally favorable approaches, new light would be shed on these tantalizing details—and who knew what else might be revealed.

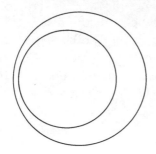

CHAPTER 6

# Satellites
# and Seeing

*When an object is once discovered by
a superior power, an inferior one will
suffice to show it afterwards.*
—*William Herschel*

ON SEPTEMBER 5, 1877, Mars
came to opposition in the constellation Aquarius, approaching
within only 35 million miles (56 million kilometers) of the
Earth.[1] Telescopes the world over were pointed in its direction
as it shone more brightly than it had for fifteen years, rivaling
even mighty Jupiter in brilliance. In the interim since the last
exceptionally favorable opposition, one or two large telescopes
had been newly turned toward the heavens, including the 26-
inch refractor of the U.S. Naval Observatory on the banks of
the Potomac in Washington, D.C., whose lens had been figured by the great telescope maker Alvan Clark of Cambridge,
Massachusetts. It had gone into operation in 1873; in charge
of it since 1875 had been Asaph Hall, onetime assistant of
William Cranch Bond and George Phillips Bond at Harvard,
who was interested in undertaking with it a careful search for
Martian satellites.

Hall and his wife Angelina had been working as school-
teachers in Shalersville, Ohio, in the 1850s when Hall decided
that he wanted to become an astronomer and applied for the
Harvard assistant's position. Though Hall had not received
much formal training (he had stayed only a year at the University of Michigan at Ann Arbor before leaving because of a lack
of funds), the elder Bond had, like Hall, been the son of a
clockmaker who had begun his astronomical career without
many advantages, and he duly hired the young man.[2]

*Figure 6.1*
*Asaph Hall (1829–1907),*
*discoverer of the satellites*
*of Mars.*

The position was not a lucrative one, and when G. P. Bond, who had been away from the observatory when Hall first arrived, met him, as Hall later recalled,

> he had a free talk with me, and found out that I had a wife, $25 in cash, and a salary of $3 a week. He told me very frankly that he thought I had better quit astronomy, for he felt sure I would starve. I laughed at this, and told him my wife and I had made up our minds that we were used to sailing close to the wind, and felt sure we would pull through.[3]

The early history of the Harvard College Observatory indeed owes much to the distinguished contributions of its devoted but poorly remunerated assistants. Among those at the observatory while Hall was there were Horace P. Tuttle, discoverer of comets and observer of nebulas, his brother Charles W. Tuttle, who later gave up astronomy for the law, and Sidney Coolidge, a grandson of Thomas Jefferson who was killed in the Civil War at the battle of Chickamauga in May 1863.[4] The last two were keen-eyed observers, especially of Saturn during the 1850s.

At the time Hall came to Harvard, its 15-inch Merz refractor was still the largest of its class in the United States and, with the 15-inch at Pulkova Observatory in Russia, the largest in the world. By the time Hall left Harvard in 1863 to take a more lucrative position at the U.S. Naval Observatory, it had been surpassed by the 18½-inch refractor of the Dearborn Observatory in Chicago, unveiled the year before, the first of the great Clark refractors.[5]

Many years previously, and not long after the erection of the 15-inch Merz refractor at Cambridge in 1847, Alvan Clark, a portrait painter and miniaturist with a studio in Cambridge, had come over to the nearby observatory to examine the new telescope's lens. He had recently become interested in telescope making in a serious way, but it seems that his first attempts had left much to be desired. W. C. Bond recorded in his diary in 1846 the receipt of a letter from Clark,

> inviting George and myself to try his new Reflecting Telescope made by himself. [H]e says he has now done his utmost to perfect it. I have been three times already when it was said to be in good order, but its performance was quite inferior—giving nothing but a round disc to a star. It is the same old story, I think; dozens of these small things have been made heretofore and the same old story told about them—thus far there is no proof of his having accomplished anything uncommon, and yet one is blamed and thought hardly of, for not extolling it as a wonderful affair.[6]

By the time of his visit to Harvard, however, Clark had given up reflectors for refractors and was expert enough in optics to be able to locate errors in the figure of the celebrated lens. Though very small, these were, he later recalled, "just enough to leave me in full possession of all the hope and courage needed to give me a start" in the exacting business of making refractors, to which he had now decided to devote himself.[7]

At first, orders were disappointingly slow. But in 1852 Clark discovered the double star 95 Ceti with one of his telescopes, a 7½-inch refractor. Among those who took notice was Dawes, who borrowed the telescope and testified to its excellence. Later he purchased this and several other Clark instruments.

The endorsement of one of the world's most esteemed observers brought Clark recognition at last, and thereafter his career advanced rapidly.

By 1862 Clark contracted to build the 18½-inch Dearborn, which became famous during testing when his son Alvan Graham Clark discovered with it the faint companion star of Sirius.[8] When the U.S. Naval Observatory announced plans for a powerful new refractor of its own, lamenting that currently its largest telescope, a 9.6-inch Merz, was surpassed by "many owned by colleges and even some owned by private individuals,"[9] Clark was the logical choice to build it. Funds for the telescope, which was to have an aperture of 26 inches, were appropriated by an act of Congress in July 1870. After obtaining suitable glass disks, Clark needed a year and a half to finish the lens. At last, in 1873, it was ready to be tested, and Clark had it mounted in a makeshift tube in the yard behind his Cambridge workshop so that it could be pointed toward the sky. Simon Newcomb, who was present on this occasion, later reminisced:

> I have had few duties which interested me more than this. The astronomer, in pursuing his work, is not often filled with those emotions which the layman feels when he hears of the wonderful power of the telescope. . . . Now, however, I was filled with the consciousness that I was looking at the stars through the most powerful telescope that had ever been pointed at the heavens, and wondered what mysteries might be unfolded. The night was of the finest, and I remember, sweeping at random, I ran upon what seemed to be a little cluster of stars, so small and faint that it could scarcely have been seen in a smaller instrument, yet so distant that the individual stars eluded even the power of this instrument. What cluster it might have been it was impossible to determine.[10]

Newcomb was in charge of the telescope when it began full operations in its dome on the banks of the Potomac in November 1873, but within two years he moved up to higher administrative responsibilities, and he was replaced in his former capacity by Hall. Hall's first order of business with the instrument was to obtain observations of the faint satellites of the outer planets Uranus and Neptune. The importance of these observations was in allowing more accurate values of the masses of these planets to be calculated. The mass of a body is known from the gravitational pull it exerts on another body. However, since the other body pulls back with equal force, in general one observes only their relative displacement, from which follows a value for the combined mass of the system rather than the mass of either body alone. The exception is when one of the bodies has a mass that is insignificant in comparison to the first, as in the case of planets with respect to the Sun or satellites with respect to planets. Then the center of mass of the system is for all practical purposes coincident with the center of the larger body, the addition of mass from the smaller body to the total being miniscule. In other words, the larger body can be taken as stationary, and its mass, $M$, follows directly from Newton's form of Kepler's third law:

$$M = 4\pi^2 r^3 / GT^2,$$

where $r$ is the mean distance of the planet from the Sun (or the satellite from the planet) and $T$ is the time required to make one revolution. A moonless planet, on the other hand, must have its mass inferred indirectly by disentangling its contribution to the disturbed motion of another planet from those of the other bodies in the solar system.

At the time, Mars fell into the moonless category. As a result, its mass was not at all accurately known. The values given for it ranged all the way from 0.09 to 0.13 times that of the Earth. While engaged in his studies of the faint Uranian and Neptunian satellites, Hall recognized that the forthcoming opposition of Mars in 1877, combined with the power of the Washington telescope, might justify a new search for Martian satellites. Moreover, the previous December he had found a brilliant white spot in the equatorial zone of Saturn, which had led to a new rotation period for the planet, that which had long been cited in the textbooks having been in error by sixteen minutes. Hall said that this made him generally skeptical of the statements in the textbooks, of which one of the most hallowed by time was, "Mars has no moons." Hall knew of William Herschel's unsuccessful search for Martian satellites in 1783, and of Heinrich d'Arrest's in 1862–64. The failure of d'Arrest, first, with J. G. Galle, to see Neptune in 1846, was particularly daunting. "The fact," Hall wrote, "that D'Arrest, who was a skillful astronomer, had searched in vain was discouraging; but remembering the power and excellence of our glass, there seemed to be a little hope left." He felt, nevertheless, that the chance of finding any satellites was very remote, and added, "I might have abandoned the search had it not been for the encouragement of my wife."[11]

HALL BEGAN the quest in early August. He was worried for a time about his ambitious assistant, Edward S. Holden (later the first director of the Lick Observatory), who was working with him at the beginning of the search and who, he feared, might upstage the discovery. As Hall confessed many years afterward in a letter to Harvard's E. C. Pickering,

> In the case of the Mars satellites there was a practical difficulty of which I could not speak in an official report. It was to get rid of my assistant. It was natural that I should wish to be alone; and by the greatest good luck Dr. Henry Draper invited him to Dobb's Ferry at the very nick of time.[12]

The first objects of Hall's scrutiny were faint stars at some distance from Mars itself, but when none of these proved to have any genuine association with Mars, lagging behind it as it pursued its motion along the ecliptic, he pushed his search closer in. Finally he found himself looking within the glare surrounding the planet. This work required special observing techniques, such as "sliding the eyepiece so as to keep the planet just outside the field of view, and then turning the eyepiece in order to pass completely around the planet."[13]

On the night of August 10, the first on which Hall attempted to examine the inner space near Mars, he found nothing, but the seeing was atrocious, and the image of the planet, seen through the boiling air of the lowlands along the Potomac, appeared "very blazing and unsteady." The next night, August 11, the search continued:

> At half past two o'clock I found a faint object on the following side and a little north of the planet, which afterward proved to be the outer satellite. I had hardly time to secure an observation of its position when fog from the Potomac River stopped the work. Cloudy weather intervened for several days. On the night of August 15, the sky

cleared up at eleven o'clock . . . but the atmosphere was
in a very bad condition, and nothing was seen of the
object, which we now know was at the time so near the
planet as to be invisible. On August 16, the object was
found again on the following side of the planet, and the
observations of that night showed that it was moving
with the planet, and, if a satellite, was near one of its
elongations. On August 17, while waiting and watching
for the outer satellite, I discovered the inner one. The
observations of the 17th and 18th put beyond doubt the
character of these objects.[14]

The satellites were named Phobos ("Fear") and Deimos
("Terror") after the two attendants of Mars mentioned in the
*Iliad*:

> So he spoke, and ordered Fear and Terror to harness
> his horses, and himself got into his shining armor.[15]

Hall kept them under surveillance through October, but by
November Mars was too far away for even the great telescope
to show them; they were not to be seen again until the next
opposition, in 1879. However, Hall had obtained enough observations to work out their orbits, from which he deduced a
value for the Martian mass of 0.1076 times that of the Earth.

THOUGH THE EXISTENCE of Hall's satellites could hardly
be denied and was in fact soon verified by other observers
using smaller telescopes, this was, ironically, to lead to a minor
controversy concerning the quality of the great telescope itself.
Some were surprised that telescopes of less than half the aperture were capable of showing the new satellites (Deimos was
glimpsed in the observatory's own 9.6-inch Merz refractor). In
addition, certain objects were reported to be visible in smaller
telescopes which resolutely defied detection in the larger one.
Otto von Struve, using the 15-inch refractor at Pulkova Observatory, had claimed the discovery of a faint companion star of
Procyon in 1873, which the Washington 26-inch persistently
failed to show (Simon Newcomb stayed up well past midnight
on one occasion looking for it, without success). Giovanni
Virginio Schiaparelli in Milan, using only an 8.6-inch refrac-

tor, made observations of surface features on Mars which also
escaped the observers at Washington. It was rumored that the
great lens had perhaps been strained under its own weight and
that this was the reason it was not performing as well as
hoped.

It was later found, however, that Struve's reputed discovery
was a mistake—the companion of Procyon was not actually
detected until John Martin Schaeberle did so in 1896 using an
even larger telescope than that at Washington, the Lick 36-
inch. As for Schiaparelli's observations, they were to remain
in dispute for many years and will be taken up in the next
chapter.

Aside from the question of the objects that the Washington
telescope *failed* to show, the other question—why the Martian
satellites detected with it now began to lie within the reach of
observers with much smaller instruments—should have posed
no mystery. In fact, it is often wondered, after an object is
once discovered, that it was not found long before. Faint objects are often reported as detectable in telescopes two or even
three times smaller than those used for their discovery.[16] Following the discovery of Saturn's inner ring (the crêpe ring) by
the Bonds in the United States and Dawes in England, for
instance, the new ring was seen by a number of observers with
small instruments. Webb mentions finding traces of it with
only a 3.7-inch refractor and adds, "anything larger, if good, is
sure to bring it out."[17] Yet if it was so easy to see, why should
its discovery have been so long delayed? In Webb's words,
"The Ring C, the crape or gauze veil, is one of the greatest
marvels of our day. How it could have escaped so long, while
far minuter details were commonly seen, is a mystery."[18]

Webb himself proposed that perhaps the ring had gradually
grown more luminous with time. There can be no doubt,
however, that the true change was not in the ring but in the
observers. The reason that the earlier observers had failed was
not that the ring did not exist or that their instruments were
incapable of showing it; it was that they did not know *how* to
see it. Charles Babbage, in *The Decline of Science in England*
(1830), notes concerning the attempts of naive observers to
make out the Fraunhofer lines in the spectrum for the first
time:

I will prepare the apparatus, and put you in such a position that Fraunhofer's dark lines shall be visible, and yet you shall look for them and not find them: after which, while you remain in the same position, I will instruct you how to see them, and you shall see them, and not merely wonder you did not see them before, but you shall find it impossible to look at the spectrum without seeing them.[19]

Babbage's words may be compared to what Dawes says concerning the occasion of William Lassell's first observation of the new Saturnian ring shortly after its discovery:

On December 2 Mr. Lassell came to see me . . . and the next night, the 3rd, being fine, I prepared to show him this novelty, which I had told him of and explained by my picture; but naturally enough, he was quite indisposed to believe it could be anything he had not seen in his far more powerful telescope. However, being thus prepared to look for it, and the observatory being darkened to give every advantage on such an object, *he was able to make it all out in a few minutes.* . . . So true are the words of Sir W. Herschel himself, "When our particular attention is once called to an object, we see things at first sight that would otherwise have escaped our notice."[20]

Another example of the same kind, also involving Saturn's rings, concerns the divisions reported from time to time in the two brighter rings (A and B). Cassini, in 1675, had found the main division, which now bears his name. At the end of the next century Laplace, in a study of the stability of the rings based on Newtonian mechanics, concluded that the pair of rings divided by Cassini's division must, if solid, be rotating around the planet to keep from collapsing into it, yet even such rings would be unstable unless they in turn were made up of a series of still narrower ringlets.[21] In fact, Laplace was too conservative; even these would be unstable, and James Clerk Maxwell later showed that the rings could only survive if they were made up of a swarm of small satellites revolving in Keplerian orbits about Saturn.[22] For our purposes it matters less that Laplace was wrong than that his work provided a fresh impetus to observers, who looked for—and in time duly found—new subdivisions in the rings. One of the first to do so was Captain Henry Kater of England's Royal Society, who in 1825, with only a 6¼-inch reflector, said that he saw (or, in fairness to him, said that he "fancied" that he saw) the outer of Cassini's two rings (ring A) broken by "numerous dark divisions, extremely close, one stronger than the rest dividing the ring almost equally." And he concluded in terms that would have gratified Laplace: "I have little doubt, from a most careful observation of some hours, that that which has been considered as the outermost ring of Saturn consists of several rings."[23] Quételet at Paris, Encke at Berlin, and De Vico at Rome also made out subdivisions in one or the other ring, apparently making Laplace's work secure.[24]

With the discovery of the semitransparent crêpe ring in 1850, however, and the seeming evanescence of the divisions of the other two rings that had been observed over the years, doubts began to arise, leading G. P. Bond to propose that the rings might be fluid. He announced the idea in 1851.[25] That same year, on October 20, Charles W. Tuttle, using the 15-inch Harvard refractor with 861×, obtained a remarkable view of the bright ring B, which he afterward described from memory as follows: "The divisions were not unlike a series of waves: the depressions corresponding to the spaces between the rings, while the summits represented the narrow bright rings themselves. The rings and the spaces between were of equal breadth."[26] There can be little question that Tuttle's perception of the rings on this occasion was influenced by the fluid ring hypothesis. Interestingly, though the fluid rings enjoyed only a brief heyday, in later years, as observers adopted other theories concerning the structure of the rings, the markings as glimpsed showed a remarkable tendency to accommodate themselves to the precise form required by theory—the degree of correspondence in the minutest details being so impressive, in some cases, that it seems impossible to account for it without assuming on the part of the observers a large dose of what Bacon called *anticipatio mentis* (mental preconception).

What all this shows is simply that the observer's selective attention is itself a very powerful factor in planetary observations. In the words of the English astronomer E. Walter Maunder, "The degree . . . to which some point attracts special attention will have much to do with the relative proportion given to it"—and, we may add, not only to the relative proportion but also to the very form itself.[27] The attention may be directed to an object by its prior discovery with a more powerful instrument, as in the case of the satellites of Mars. No less often it seems to be directed by some theoretical advance: Would the minor subdivisions in Saturn's rings likely have come to light so soon but for the prior calculations of Laplace?

William James, in his discussion of attention in *The Principles of Psychology* (1890), writes: "Most people would say that a sensation *attended to* becomes stronger than it otherwise would be," and he illustrates this with several examples:

> Every artist knows how he can make a scene before his eyes appear warmer or colder in color, according to the way he sets his attention. If for warm, he soon begins to *see* the red color start out of everything; if for cold, the blue. Similarly in listening for certain notes in a chord, or overtones in a musical sound, the one we attend to sounds probably a little more loud as well as more emphatic than it did before. When we mentally break a series of monotonous strokes into a rhythm, by accentuating every second or third one, etc., the stroke on which the stress of attention is laid seems to become stronger as well as more emphatic. The increased visibility of optical after-images . . . which close attention brings about, can hardly be interpreted otherwise than as a real strengthening of the retinal sensations themselves.[28]

The founder of the first laboratory for experimental psychology, Wilhelm Wundt, had noted some years previously the effect of attention on shortening reaction times in his subjects and had concluded: "We cannot well explain these results otherwise than by assuming that the strain of the attention towards the impression we expect coexists with a preparatory innervation of the motor centre for the reaction, which innervation the slightest shock then suffices to turn into an actual discharge."[29] James accepts this explanation and expands on it, proposing that the "lying in wait for the impressions, and the preparation to react, consist of nothing but the anticipatory imagination of what the impressions or the reactions are to be." Later he notes,

Where the stimulus is unknown and the reaction undetermined, time is lost, because no stable image can under such circumstances be formed in advance. But where both nature and time of signal and reaction are foretold, so completely does the expectant attention consist in premonitory imagination that, as we have seen, it may mimic the intensity of reality, or at any rate produce reality's motor effects.[30]

James also indicated that "it is for this reason that men have no eyes but for those aspects of things which they have already been taught to discern. Any one of us can notice a phenomenon after it has once been pointed out, which not one in ten thousand could ever have discovered for himself."[31]

Labeling, in James's view, turns out to be but another form of preperceiving, and he adds that "we commonly see only those things which have been labelled for us." It follows that the question of how the labeling is done is far from arbitrary. We have already seen the degree to which certain labels, such as "seas" and "lands" applied to regions of the Moon, gave rise to definite expectations of what was to be found there—gave rise, indeed, to a prejudice that was eventually overcome only with considerable difficulty. With this in mind, we now turn to the work of the Italian astronomer Giovanni Virginio Schiaparelli, who in the same year that saw Hall's discovery of the satellites set himself the task of devising a new and—so he thought—enlightened set of labels for the surface features of Mars. Certainly it might be said in retrospect that these labels were to give astronomers for some time to come a new way of preperceiving the planet.

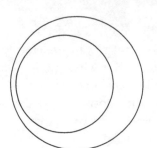

CHAPTER 7

# Schiaparelli

*Fraudful Mercury.*
　　　　　—*Sophocles,* Philoctetes

GIOVANNI VIRGINIO SCHIA-
PARELLI was born on March 14, 1835, in the little town of
Savigliano in the Piedmont region of northwestern Italy, not
far from the French border. The town lies at the foot of the
Alps and is overlooked by an ancient Benedictine Abbey. Like
many border regions, with their shifting boundaries, the Piedmont tends to be polyglot, and the young Schiaparelli developed a remarkable facility in languages; he was to master
French, German, English, Latin, Greek, and Arabic—an accomplishment that opened up to him a cosmopolitan correspondence with the astronomers of his day and that later greatly
assisted the researches into the history of astronomy that became a consolation in his old age.[1] So much we may trace, in
the development of his character and interests, to the geographical setting of his youth; but for his most celebrated
discovery, that of the "canals" of Mars, it would have been
more fitting had he passed his formative years not in Piedmont
but on the other side of the peninsula—in Venice.

His interest in astronomy was first aroused by a learned
priest of Savigliano, Paolo Dovo, of whom Schiaparelli later
wrote, "he was a man of gold, a great lover of astronomy, and
one whose image could never be erased from the memory of
those who had known him."[2] Yet at first Schiaparelli had plans
to stake out a different career. On leaving his native town for
the university in nearby Turin, the future discoverer of the
"canals" of Mars intended to become a civil architect and an
engineer—specifically (and not a little ironically, given the sequel) a hydraulic engineer. Schiaparelli excelled in mathematics, and following his graduation with honors he taught
elementary mathematics for a brief time at a Gymnasium in
Turin. He was apparently trying to make up his mind what he
really wanted to do with his life. The decision came a short

Figure 7.1
G. V. Schiaparelli (1835–1910)
at the height of his powers as a
planetary observer.

time later: "Without taking into account my almost absolute poverty," he afterward recalled, "I formed the project of devoting myself to astronomy, which was not done without much opposition on the part of my parents."[3] Thus in February 1857 Schiaparelli set out under the sponsorship of the Piedmontese government for the Berlin Observatory, then under the directorship of J. F. Encke. However, after a period in Berlin he became dissatisfied with the training, noting that "the teaching of Encke was devoted rather to detail than to observation."[4] He went thence to Pulkova Observatory, at the time headed by Otto von Struve, and on returning to Italy in 1860 obtained a post as *secondo astronomo* under Carlini at the Brera Observatory in Milan. The following year he discovered the asteroid Hesperia, the sixty-ninth found in the space between Mars and Jupiter, and when Carlini died in 1862, became his successor in the directorship at the age of only twenty-seven. In 1865 he married Maria Comotti, with whom he had two sons and three daughters.

IN AUGUST 1862, Schiaparelli carried out a series of observations of the large comet of that year (1862 III) that had been discovered by the American observers Horace Tuttle and Lewis Swift. Telescopically, fountainlike jets of luminous matter appeared to stream from the starlike nucleus. The impression seemed inescapable that the comet was in the process of disintegrating. Similar appearances had been observed in other large comets and had led Father Secchi at the Vatican Observatory in Rome to propose his fountain model: upon heating of the nucleus by the Sun, material was supposed to be shot up from the comet and swept backward in parabolic streamers rather like those produced when water is shot up in a fountain. Secchi had further speculated that perhaps this debris might have some relationship to meteors, which are produced when tiny particles encounter the Earth's atmosphere and are burned up. Schiaparelli developed the idea in a series of lectures given to the Royal Institute of Lombardy, afterwards published as "Le Stelle Cadenti" (The falling stars).[5] The work was hailed by Sir Norman Lockyer as one of the greatest contributions to astronomical literature produced in the nineteenth century.[6] "The meteoritic currents," Schiaparelli said, "are the products of the dissolution of comets and consist of minute particles which certain comets have abandoned along their orbits." Schiaparelli's achievement was in going on to confirm this idea by showing that an actual meteor swarm—the Leonids, which had produced spectacular showers in 1799, 1833, and 1866—followed the orbit of Comet Tempel (the same conclusion was reached in an independent investigation by France's U.J.J. Leverrier). Whenever the Earth crossed the orbit of the comet, it encountered the swarming debris that had been left behind, and a meteor shower occurred. At about the same time, in a series of letters to Secchi published in the *Bullettino Meteorologico dell'Osservatòrio del Collegio Romano* in 1866, Schiaparelli showed that the most brilliant annual shower, the August Perseids, was associated with the Comet of 1862 itself.[7] His meteor work won for Schiaparelli the prestigious Lalande prize of the French Academy; he would receive it again for his later work on the rotations of Mercury and Venus.

SCHIAPARELLI'S YOUTHFUL WORK on meteors was perhaps his most solid contribution to astronomy, and it remains basically unassailed to this day. He became even more celebrated, however, for his work on the planets, to which he turned in his early forties when the Brera Observatory finally obtained an instrument suitable for the purpose, an 8.6-inch Merz refractor, which went into operation in 1875.

Two years later, of course, he made his famous observations of Mars. That same year he also began an investigation into the rotations of the inner planets. At the time, all the planets from Mercury through Mars were widely believed to rotate in an Earthlike period of about 24 hours. Already in the seventeenth century, Huygens, from his observations of the spots of Mars, had concluded that "the rotation of Mars seems to take 24 terrestrial hours like that of the Earth." Cassini's more careful investigation later amended this to 24 hours, 40 minutes. The less definite spots of Venus had suggested to Cassini a similar rotation period for that planet, as we have seen; and Schröter and De Vico at the beginning of the nineteenth century had concurred. Meanwhile, Schröter had also proposed a period of 24 hours for difficult-to-observe Mercury, a result which, in the absence of any better estimate, stood alone.

In 1877, Schiaparelli noted two well-defined bright spots near the cusp of Venus, which led him to suspect a long rotation, but subsequently he failed to find anything so definite on the planet—only the typical nebulous dark shadings, which he regarded as "dangerous to make use of for the rotation."[8] Most of the markings that other observers had seen he regarded as illusory; nevertheless, taking the observational record as a whole into account, he concluded in 1890 that it was probable that the planet had an isosynchronous, or "captured," rotation—that is, that it rotated in a period equal to that of its revolution, 225 Earth days. In other words, Venus must keep the same face always turned toward the Sun.[9] Another series of observations, made in 1895, was confirmatory of the first.

If the results on Venus were still tentative, owing to the indistinctness of its features, Schiaparelli's work on Mercury seemed much more firm.[10] This planet had never been easy to observe because, being the innermost, it always remains near the Sun. While it comes to greatest elongation about once every other month, swinging far enough away from the Sun so that it can be seen without too much difficulty with the naked eye as a morning or evening "star," not all of these presentations are equally favorable. In the northern latitudes the best of them occurs in the spring for Mercury as an evening star and in the autumn for Mercury as a morning star, because the ecliptic stands most nearly perpendicular to the horizon at those times. Yet even under optimal circumstances, Mercury never sets more than about two hours after the Sun or rises more than about two hours before it. This means that even when it is shining brightly as seen with the naked eye, its light is passing through the thickest part of the Earth's mantle of air; its unplanetlike twinkling at these times—which earned for it the name of *Stilbon,* the "scintillating one," among the ancients—bodes ill for the telescopic observer, who must then try to make do with an image that is hopelessly confused and distorted for observations of finer details. Schiaparelli hit upon a splendid idea for getting around this problem. One would simply have to observe the planet during the daylight hours, when it was higher up. Since Schiaparelli's telescope was equipped with setting circles, it was not difficult to point it to the right part of the sky. He made a few tests of the technique in June 1881, was encouraged by the results, and in the following year took up the problem of Mercury's rotation in a serious way.

Unfortunately, even when higher in the sky, Mercury remains a difficult object. Its diameter is only one and a half times that of the Moon, making it the smallest planet except Pluto, yet it never approaches within less than 50 million miles (80 million kilometers) of the Earth. At the time that its disk is largest, its phase is either "new" or a thin crescent, so that little of its surface can be examined. As the phase increases, so does the distance, until when the phase has finally become full the planet has moved all the way to the other side of the Sun. Schiaparelli made a concerted effort to keep the planet under as continuous observation as possible. Some idea of his industry can be gleaned from the record of the dates that he observed the planet during 1882 alone: February 4–10, March 31–April 28, May 24–31, August 5–21, and September 19–30. Naturally, most of these dates corresponded to times of the planet's elongations, though in August the planet was near superior conjunction—Schiaparelli followed it to within $3^{1}/_{2}°$ of the Sun, when its phase was nearly full and its disk only $4''$ of arc in diameter. The next year, 1883, he observed the planet almost as intensively, but from 1884 on the number of his observations began to decline somewhat.

Schiaparelli was able to identify markings on Mercury which, he said, usually appeared "in the form of extremely faint streaks, which under the usual conditions of observations can be made out only with greatest effort and attention."[11]

Only during the hot Italian summer was the image generally
totally unusable, the atmosphere being too agitated then for
any markings to be made out. Incidentally, the summers made
work difficult for other reasons as well. As Schiaparelli wrote
to the English astronomer E. B. Knobel, "The heat is great,
and the processes of life and work proceed at the lowest possible pressure. I am incapable of doing anything." And he
added: "To sleep and not to wake again would be what would
please me best now."[12]

Schröter had reported that the markings on Mercury appeared in the same positions when observed in the twilight
periods on successive days and thus had come to his conclusion that the rotation period must be about 24 hours. However, Schiaparelli pointed out that other periods were equally
compatible with this observation. Having liberated himself
from the requirement of studying the planet only in the brief
twilight periods, he was able to keep the markings under observation for several hours at a time; what he found was that
they remained perfectly stationary over this interval. Had Mercury rotated in a period as short as 24 hours, a noticeable shift
in them should have taken place; since it did not, the period
had to be significantly longer than 24 hours.

So far so good. Yet if Schröter had failed to consider the
problem more fully because of the analogy he saw to the
Earth, Schiaparelli was led by a different avenue to premature
closure. His observations were not long underway before he
had reached the conclusion that the rotation was captured
with respect to the Sun. This was, in fact, logical enough. The
Moon always keeps the same face toward the Earth; Saturn's
satellite Iapetus, discovered by Cassini in 1671 and noted by
him to be two magnitudes brighter when farthest west of
Saturn than when farthest east, was shown by Sir William
Herschel to be a similar case. Herschel wrote presciently in
1792: "We may conjecture that probably most of the satellites
are governed by the same law; especially if it be founded on
such a construction of their figure as makes them more pon-
derous toward their primary planets."[13] This state of affairs
was satisfactorily accounted for by the theory of "tidal fric-
tion," which was first hinted at by Immanuel Kant in the
eighteenth century and was revived by others in the middle of

the nineteenth, a particularly thorough study of the subject in the case of the Moon being carried out by Sir George Howard Darwin (the great naturalist's son). Given the immense power of the solar tidal forces at Mercury's distance from the Sun, it was certainly reasonable to conclude that the same process had led to a captured rotation there. Indeed, such a result had even been predicted by the American astronomer Daniel Kirkwood as early as 1865.[14]

Despite Kirkwood's previous suggestion to the same effect, E. M. Antoniadi was quick to point out that "this in no way lessens the credit due to Schiaparelli, bearing in mind the smallness of the telescopes which he used. Indeed, his work on Mercury constitutes the most beautiful of all the telescopic discoveries made by the great Italian astronomer."[15] One wonders whether Schiaparelli's keenness of sight may just possibly have been assisted by an "anticipatory imagination" provided by the previous theoretical work. Be that as it may, once he had committed himself to the notion of the captured rotation it proved impossible to get past it, even when his observations furnished him some impressive hints that in fact things might be otherwise.

There was a definite drift of the markings across the disk, but this was easily explained away as a result of libration, familiar in the case of the Moon and owing to two motions getting out of step with one another. Because of the conservation of angular momentum, the spinning of the Moon on its axis is constant. At the same time, as required by Kepler's second law, the Moon as it moves in its elliptical orbit around the Earth must travel with greater velocity when closest to the Earth (at perigee) than when farthest from the Earth (at apogee). Near perigee the orbital motion gets ahead of the axial spin; near apogee it falls behind. The result is a slight rocking back and forth, which allows Earthbound observers to peek over the limb to regions that would otherwise lie hidden on the far side. The same situation was only to be expected to apply to Mercury in its highly elliptical path around the Sun. In fact, because its orbit is so eccentric, the libration would be expected to be of considerable amplitude. This allowed for a fair degree of variation in the observed positions of the surface features. Admittedly there was something strange about the supposed libration of Mercury: as Schiaparelli tracked the markings from day to day, they always shifted in the direction that would have been expected on the basis of a regular rotation of the planet. They were never caught in the act of swinging back again.

Yet even assuming a marked libration, Schiaparelli still found, as he admitted to the English observer W. F. Denning in 1882, that the markings on Mercury were "extremely variable." Sometimes he found them "partially or totally obscured," and he noted that the planet "also has some brilliant spots which change their position."[16] Recall that at this point he had had the planet under observation for only a year. Perhaps the variability of the markings was what caused him to delay publishing his conclusions right away; it was presumably less because he questioned the rotation period than because he wanted to confirm his suspicion that the observed changes indicated the presence of a considerable atmosphere around the planet. Beginning in 1886 he was able to make use of a larger telescope, an 18-inch refractor. With this he observed Mercury for three more years, recording his final observations on April 19 and 20, 1889. At long last he was ready to publish.

He wrote in his 1889 monograph *Sulla Rotazione di Mercurio* (On the rotation of Mercury) that the dark spots were "sometimes more clearly visible, sometimes less distinct," and that they "invariably disappeared at the neighborhood of the limb. . . . But it often happens that in the central region of the disk, the spots are weakened to a greater or lesser degree, becoming completely or partially invisible, and remaining affected for several days."[17] Since the captured rotation had, in his view, been put beyond doubt, he had no choice but to account for these changes as due to "veils" which he described as "more or less opaque condensations produced in the atmosphere of Mercury, which from afar presents aspects analogous to those which our Earth would show from a similar distance."[18]

Schiaparelli seems to have had no conceptual difficulty with the fact that his observations had led him to suppose Mercury to be a kind of planetary equivalent of the centaur, half man and half horse. Schiaparelli's Mercury was half like the Moon, half like the Earth: like the Moon in its rotation, like the Earth in its atmosphere. Indeed, Schiaparelli did not even see any reason to rule out its habitability. Father Secchi—whom he quoted with approval on the subject of extraterrestrial life in one of his later publications about Mars—had written: "In our opinion, it seems absurd to regard the vast regions [of the universe] as hardly inhabited deserts; rather, they must be richly populated with beings intelligent and rational, capable of knowing, honoring, and loving their Creator."[19] Perhaps one can find a hint of this view of Secchi's in Schiaparelli's concept of the planet. Fontenelle had made Mercury the home of hot-tempered Latins, because of its proximity to the Sun; Schiaparelli did not go this far, but he pointed out that conditions might actually be rather temperate there, owing to winds continually blowing air from the night onto the day side. As for the spots, he declared that the difficulty in observing them made opinions about them uncertain. They might be similar, he said, to the spots on the Moon. On the other hand, "if anyone, taking into account the fact that there exists an atmosphere upon Mercury capable of condensation and perhaps also of precipitation, should hold the opinion that there was something in those dark spots analogous to our seas, I do not think a conclusive argument to the contrary could be advanced."[20] This negative way of stating a position is, as we shall see, a Schiaparellian trademark.

FOLLOWING THE PUBLICATION of Schiaparelli's results, other observers took up the planet. In 1890, Henri Perrotin, with the 30-inch refractor of the Nice Observatory on the French Riviera, became the first to confirm the captured rotation period. Then in 1896–97 Percival Lowell, observing with a 24-inch refractor both at Flagstaff, Arizona, and at Tacubaya, Mexico, made out linear markings on the planet (Fig. 7.2), of which he wrote: "So visible were they at all times—for the air had to be distinctly bad to obliterate them—that observations had not been made more than a day or two before the rotation period of the planet was patent."[21] He also noted the libration, which he called "a definiteness in the proof of a really surprising kind."[22] It might have seemed less definite had Lowell recognized the significance of the fact that the presumed libration was always seen to be taking place in the same direction. Expectation created the illusion of libration. Only on the matter of Mercury's supposed atmosphere did Lowell beg to differ with the great Italian observer: he

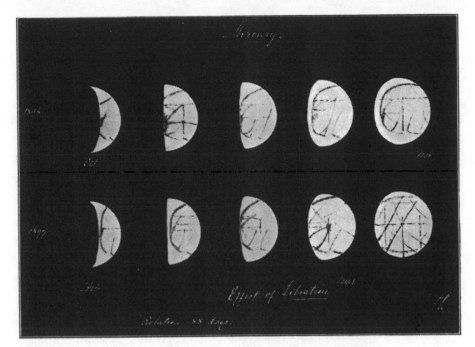

Figure 7.2
Libration of Mercury, according to Percival Lowell. These drawings were made in 1896–97, the first year of operation for Lowell's 24-inch refractor. They show the supposed lurching back and forth of the planet that was cited by Lowell as evidence of the 88-day rotation proposed by Schiaparelli in 1889. It is now known, however, that the longitude of the central meridian at the elongation of October 23, 1896 (the half-phase view in the top series), was some 100° different from that at the elongation of February 15, 1897 (the half-phase view in the bottom series). It is therefore difficult to know what to make of the similarity of the markings depicted in the two views, but the presumption is that the markings were completely illusory.

failed to find evidence of any and added that in any case, "theory informs that it should be nearly, if not wholly, absent" on a planet so small and so near the Sun.[23]

W. F. Denning, though using only a 10-inch reflector, did at least observe the planet over a very long time. Unlike Lowell, he confirmed the clouds but not the rotation. Working entirely in the twilight periods, he found the features clearly recognizable from one night to the next but felt that they had slightly altered their positions. "Their movement on the disk," he wrote, "appeared inconsistent with a very long rotation." He therefore went back practically to the rotation period claimed by Schröter, though increasing it to 24 hours, 42 minutes. The presence of a dense Mercurial atmosphere was invoked as the only way of accounting for his experience that the "faint, irregularly shaped, dusky spots and white areas" he saw on Mercury had no duration, as evidenced by the fact that they were, as he put it, "not always easy of identification over long periods."[24]

Various other observers investigated the question, but none was more highly respected than Eugène Marie Antoniadi. He began a careful scrutiny of the planet in 1924 with the 32³/₄-inch refractor of the Meudon Observatory, near Paris, following Schiaparelli's method of observing the planet by daylight. "Putting aside earlier work," he wrote, "I have undertaken research which is absolutely independent, and on my drawings I have recorded only the patches which were seen with certainty."[25] He observed in 1924, 1927, 1928, and 1929, and he concluded that Schiaparelli had been correct both about the rotation and the clouds (Fig. 7.3).

Antoniadi believed that he had recognized the same markings often enough in the same positions on the planet to be sure of the rotation. However, like Schiaparelli he found the markings to be variable, noting that

> often they were feeble to the point of being nearly invisible, but on one day, when Mercury was a few degrees from the Sun, they appeared at least as dark as the Syrtis Major and the Sinus Sabaeus on Mars. These differences explain why one observer may describe the patches as pale, whereas another may call them quite dark.[26]

There also appeared to be local obscurations. For instance, the horn-shaped dark area that he named Solitudo Criophori was often seen as very dark, and he identified it with a marking conspicuously displayed on Schiaparelli's map. However, Antoniadi found that it frequently appeared to be veiled, and wrote: "It appears to be covered by dusty clouds more often than any other greyish area."[27] He noted similar changes in other areas, so that he believed he had no choice but to conclude that the planet's clouds were "much more frequent and obliterating than those of Mars."[28] Schiaparelli seemed entirely vindicated.

THE TRUE ROTATION of Mercury was discovered in 1965 by radio astronomers. It is not equal to the period of revolution of 88 days, as Schiaparelli had maintained, but to exactly two-thirds of this, 58.65 days. At once it was asked how the visual observers could have been so far wrong—not only Schiaparelli and Antoniadi but also later observers of equal skill, notably Bernard Lyot and Audouin Dollfus, who studied the planet extensively during the 1940s and 1950s with the 24-inch refractor at the Pic du Midi Observatory in the French Pyrenees. On the basis of their work, they concluded that Mercury's orbital period and the planet's rotation on its axis coincided to within an accuracy of one part in a thousand.[29]

The 58.65-day period is the length of one "sidereal day" on Mercury—its rotation period as measured with respect to the stars. One "solar day," the time between successive sunrises at any point on the planet, is 176 days. However, as first pointed out by Clark R. Chapman and Dale P. Cruikshank, the figure of 176 days is also significant because it is $1\frac{1}{2} \times 117$ days, the interval between successive evening or morning elongations (the so-called synodic period).[30] Following a favorable autumn elongation, when Mercury is a morning star, the same face will be lit up by the Sun 176 days later—when Mercury is appearing as an evening star. On the latter occasion, observers on Earth will actually be looking at the opposite hemisphere of the planet, but because of the phase, the fact will go undisclosed.

If, now, the planet is followed to the next favorable morning elongation, in the autumn, it will have rotated six times, and

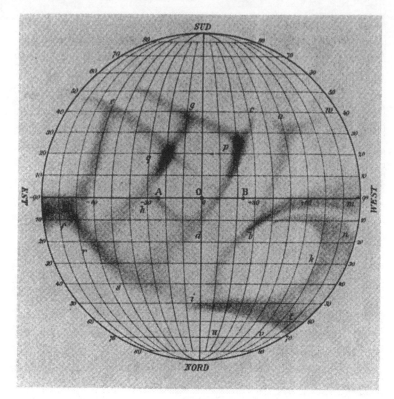

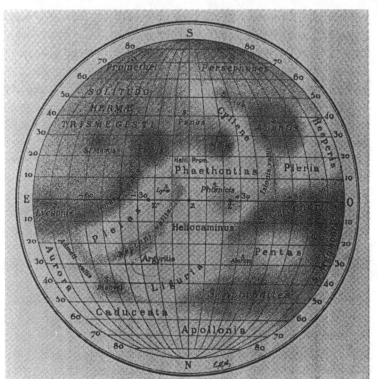

*Figure 7.3*
*Maps of Mercury by Schiaparelli,*
*1889 (left), and by Antoniadi,*
*1933 (right). Though the markings*
*are depicted as rather narrower by*
*Schiaparelli—an effect that can*
*be satisfactorily accounted for on*
*the basis of diffraction, because*
*Schiaparelli used smaller telescopes*
*than Antoniadi—the overall*
*correspondence between the two*
*maps is very striking.*

very nearly the same illuminated face will be turned toward the Earth as was visible the year before. Since the most intensive periods of observation tend to be around the times of these most favorable elongations, astronomers, seeing more or less the same features year after year and not suspecting any correlation between Mercury's motion and that of the Earth, would have had some grounds for concluding that the planet had an isosynchronous rotation.

This synchronization of opportune observing "windows" with the presentation of the same face of the planet makes up what has been called the "stroboscope illusion," and no doubt it played an important role in deceiving the observers of Mercury. Actually, though, the synchronism is by no means exact. The time between the successive favorable elongations in the autumn is only about 348 days, four days less than the time required for six rotations (6 × 58.65 = 365 days). Thus, new features will gradually roll onto the disk. Also, the 348-day period is 17 days less than one Earth year. The result is that the dates of elongation tend to creep forward from year to year, and gradually one cycle of elongations displaces another as the most favorable. If one observes long enough, the whole surface of the planet will eventually be presented for review, even if one only observes for a few days around the most favorable date each year.

The question remains, What was happening when Schiaparelli was watching and drawing Mercury in the 1880s? I have calculated the dates of Mercury's most favorable elongations west of the Sun (as he saw details better at the morning than at the evening elongations) and the approximate longitude of the central meridian for several years during this period so that the reader can see the pattern. (Following the *Mariner 10* spacecraft mission to Mercury in 1974, longitudes on Mercury have been defined so that the 20° meridian passes through the crater Hun Kal.) The dates and corresponding longitudes are:

| | | |
|---|---|---|
| 1883 | October 21 | 142° |
| 1885 | September 15 | 124° |
| 1887 | August 16 | 131° |
| 1889 | July 11 | 113° |
| 1889 | October 30 | 325° |

In terms of what has just been described, one cycle of favorable elongations was giving way to another in 1889, and the hemisphere that would be displayed for the next seven years or so was displaced a good part of the way around the planet from that of the previous cycle. Unfortunately, Schiaparelli gave up observing Mercury in April 1889.

Of course, Schiaparelli did not observe only during the most favorable elongations; as we have seen, he made it a point to try to keep the planet under as continuous scrutiny as possible. Indeed, he observed it on various occasions when different areas of its surface were in view. Yet the planet was, after all, a difficult object; its disk was small, and the features on it were hardly conspicuous. He wrote on August 7, 1882, that "all the delicate streaks are of the color of light brown melting into rose, whose nuances are only slightly different from the background, and very difficult to distinguish."[31] Again, on April 25, 1883, he remarked, "Do not deceive yourself, for the figures as represented are in part subjective, and depend on the mode with which the eye tends to group together the complicated lines and stains," and a year later he recorded, in words that would anticipate some of the later revelations of fine structure in the Martian surface, "[I] have the impression, when the seeing with the instrument becomes very steady, that all the appearances are resolved into very fine formations."[32] One had to expect that such delicate features, visible on one occasion, might well be invisible on another, less favorable one. Thus, one could indulge in a certain degree of postselective discretion in interpreting the observations to fit one's scheme. Not only that, but what was actually a different feature might, under the circumstances, be unconsciously represented in such a way as to agree with one's preconception of it; for, as Schiaparelli admitted, "the want of definite edges . . . always leaves room for a certain choice" in where one puts the boundaries in a drawing.[33]

Then there was the whole question of the clouds. Generally speaking, there was nothing that proved more problematical for visual observers of the Moon and the planets than Schröter's old project of establishing the occurrence of change. In addition to the long-term changes in Mercury that Denning referred to, both Schiaparelli and Antoniadi claimed to have seen changes in features occurring over a period of a few days. If this was the basis on which the existence of clouds was initially established, then one can only say that the evidence was indeed dubious. Changes in the visibility of planetary surface features are notorious for often being due to changes in the transparency of the Earth's own atmosphere rather than in real or assumed changes in the planet being observed. As E. Walter Maunder wrote in 1894, "When the difference of tone in . . . markings is but small, but a little defect in the transparency and steadiness of our own atmosphere will be sufficient to render them indistinguishable."[34] The existence of clouds, fogs, and atmospheric veils over certain lunar features was inferred from changes in the telescopic visibility of minute pits; in the case of Mars, inundations of continents by seas were deduced from apparent changes in the boundaries between these regions. The visual observers relied too heavily on such evidence. In the case of Mercury, the impression of clouds presumably rested at first on equally nebulous grounds, but once clouds were accepted they became an extremely convenient "fudge factor." They destroyed the "cognitive dissonance" that might have emerged from the unrecognizability of features at intervals and prevented Schiaparelli, at least, from giving consideration to other possible explanations for the variability of the Mercurial spots that he described to Denning as early as 1882.

As for the later observers of Mercury, their failure to recognize Schiaparelli's mistake underscores a fundamental problem in observational science. Lowell said that the rotation period of the planet was "patent" after only a day or two of observation. This points out the almost unavoidable tendency of later observers to take intellectual shortcuts when tackling a problem that had seemingly been dealt with thoroughly by a predecessor, particularly a distinguished predecessor like Schiaparelli. Lowell observed the planet briefly, reached his conclusions, and never returned to it. Only Denning, who, perhaps significantly, started his work on Mercury before Schiaparelli announced his results, seems to have recognized that the daily drift of the features could not be merely a result of libration; had he interpreted his own observations in terms of a long period rather than a short one, he would have

arrived at a result of the right order. But his telescope was too
small to allow him to see the features distinctly, and he could
only observe in the twilight periods. All he grasped was that
the features did seem to change over time.

By the time Antoniadi started to observe the planet, the
cycle of most favorable elongations was such that very nearly
the same configurations of the surface that Schiaparelli had
mapped were presented to view. There was not only one
"stroboscope illusion," discussed above, but a secondary one
operating as well. Observers were doubly damned by the
strange coincidences in the periods. In effect, Antoniadi's
study was merely a repetition of Schiaparelli's earlier one, as is
immediately apparent from the similarity of their charts (due
allowance being made for the inevitable differences in drawing style, as well as for the fact that Schiaparelli, observing
with his smaller telescope, would have observed the dark areas
as narrower than Antoniadi with his larger one—exactly as
would be expected on the basis of diffraction).

No doubt the case of Mercury's rotation nicely demon-
strates what we have had reason to point out before: once a
definite expectation is established, it is inevitable that one will
see something of what one expects; this reinforces and refines
one's expectations in a continuing process until finally one
is seeing an exact and detailed—but ultimately fictitious—
picture. Schiaparelli's work is a remarkable case study in autosuggestion; though proclaiming himself "independent," Antoniadi seems instead to show, in the domination of his views
by Schiaparelli's, something of that yielding of the will to
another that one sees in subjects under hypnosis. So definitive
was Antoniadi's work regarded, in turn, that later observers
contented themselves with making drawings that are often, in
the words of Chapman and Cruikshank, "subconscious reproductions of Antoniadi's chart."[35]

SCHIAPARELLI'S WORK on Mercury was once praised as
an outstanding example of observational method and skill.
Stock in Schiaparelli has fallen considerably in the last few
decades. His ideas about Mercury were disproved in 1965,
and by coincidence, the same year also saw the final disman-
tling of another of his most famous "discoveries," that of the

"canals" of Mars, by the *Mariner 4* spacecraft, which secured the first close-up pictures of the planet. Yet whereas almost no one before 1965 suspected that Schiaparelli had been wrong about Mercury, the "canals" were regarded with considerable skepticism even in his own lifetime.

Well-versed in the classics as he was, Schiaparelli might have appreciated the irony in the fact that his fame as the "discoverer" of Mercury's rotation—the achievement for which he was awarded his second Lalande medal—was to be "stolen" from him by the planet named for the patron god of thieves: this Machiavellian planet, seeming to show always one face publicly but hiding another and quite different face. By all accounts, Schiaparelli himself seems to have been a bit Machiavellian. It is not surprising to find him serving in his later years as a senator from Piedmont in the recently founded Kingdom of Italy; in all probability he felt entirely comfortable with the intrigues of Italian politics, for in his astronomical work he sometimes seems to have carried to an extreme the kind of evasiveness more customarily seen in diplomatic circles.

Nowhere is Schiaparelli's quicksilver slipperiness more in evidence than in the matter of the famous Martian "canals," which he first recorded in 1877. When in 1893 he was finally willing to grant that his "canals" were probably such in fact as well as in name, "destined for the passage of liquid . . . and constituting for [the planet] a true hydrographic system," he satisfied those on both sides of the question as to their significance by proposing that, while he himself believed them to be geologic in origin, he could personally think of no mechanism that could explain them—so that, as Percival Lowell remarked, "the suggestion is, properly speaking, not a theory."[36] On the other hand, while not openly endorsing the idea of their artificial origin, which their geometrical character had suggested to some, he refused to "combat the supposition," which, he said, "includes nothing impossible." As we have seen, he had stated his views about Mercury in a similarly negative sense: "If anyone . . . should hold the opinion that there was something in those dark spots analogous to our seas, I do not think that a conclusive argument to the contrary could be advanced."[37]

One may, perhaps, see in this an Olympian unwillingness to commit himself prematurely to any interpretation. He arued eloquently (in terms not a little reminiscent of Beer and Mädler) that "for us, who know so little of the physical state of Mars, and nothing of its organic world, the great liberty of possible supposition renders arbitrary all explanations of this sort, and constitutes the gravest obstacle to the acquisition of well founded notions."[38] Or one may see it, rather, as only a careful hedging of his bet, inferring from what he does not say the probable region of his sympathy— for the door he refused to shut he ought to have known would thereby be left open all the wider for others to rush through, as indeed they did. In the case of Mercury there seem to have been few takers, but in that of Mars the Frenchman Camille Flammarion and the Bostonian Percival Lowell were only the most notable of those who rushed in. Of Lowell, E. S. Holden was later to write, borrowing a phrase from Kipling, that he was

> Hanging like a reckless seraphim
> On the reins of red-maned Mars.[39]

Schiaparelli himself was anything but reckless. Yet like Mercury itself, which hid one face from view, he sometimes gives hints of a side that might not have been quite so conservative as the one he showed publicly.[40] One sometimes suspects that he might really have been on the side of the Martians without admitting it—particularly as he seems to have accepted, to the bitter end, observations whose peculiarity admitted hardly any possible natural explanation. This contradiction—the willingness to accept the genuineness of phenomena that hardly seemed explicable on the basis of any hypothesis except that of intelligent handiwork, and his final unwillingness to embrace that hypothesis—made him a frustrating figure for some of his contemporaries. It seemed to them, as it still does to us, as if he were somehow trying to have it both ways.

As much as one tries to resolve this apparent contradiction by finding out what Schiaparelli "really thought," which in itself belies the difficulty of believing that it was really what he professed, in the final analysis his agnosticism seems sincere. One has the choice of accepting as his view what he repeated again and again with definiteness—that the "canals," though undoubtedly quite real, were no less uncertain in their interpretation—or of making more than perhaps is justified of his various pronouncements on the other side, always rather facetiously stated or couched in careful phrases of half-denial. But whatever one decides, the man remains finally ambiguous. This is, however, perhaps only fitting for one who was among the greatest of planetary observers: for in the telescope the planets themselves are shiftingly ambiguous and seldom divulge their secrets openly, but rather vouchsafe them in veiled hints and suggestions.

SCHIAPARELLI INITIALLY UNDERTOOK his observations of Mars in 1877 in order to draw up an improved map of the planet. As a onetime civil architect and hydraulic engineer (or at any rate as one who had received training in those disciplines), he perhaps not so surprisingly came to represent the planet in the form of a diagram or "plan view" of its surface, a seemingly innocuous circumstance that was, as we shall see, not without its implications. Webb would speak, accurately enough, of his "micrometric vision."[41] Indeed, his map was founded on sixty-two positions on the surface precisely measured with the micrometer—an instrument with which he had become proficient through measuring double stars, of which measurements he was to amass some 11,000 by 1899. His discussion of the methods used to make the measures of the Martian features the basis of his map is admirably detailed and thorough. Moreover, having determined to make a fresh start on Martian cartography, Schiaparelli decided to invent a new system of nomenclature as well. By century's end, Martian nomenclature was indeed in a confusing state, with a multitude of competing and all more or less unsatisfactory schemes in use. Besides Proctor's, introduced on his 1867 map, there were also available a variant on this advocated by another Englishman, Nathaniel Green, and a somewhat less chauvinistic system, albeit one along similar lines, proposed by the Frenchman Camille Flammarion. Schiaparelli approached the problem by entirely abandoning the method used by his rivals of naming features after past observers of the planet and instead decided to give the Martian features geographical names (in the same way that

Hevelius had rejected Langrenus's personality-based scheme for the moon because of the jealousies he feared it would arouse and had constructed a geographical scheme in its place).

Inveterate classicist that he was, Schiaparelli drew on the geography of the ancient world for his names. In the east of his map was Solis Lacus, the Lake of the Sun, the legendary place where the Sun began its journey through the heavens each day, and next to it, appropriately enough, was Aurorae Sinus, the Bay of Dawn. Proceeding westward, there was Margaritifer Sinus (the Pearl-bearing gulf, the old name for the coast of India), Syrtis Major (the Gulf of Sidra), and Mare Tyrrhenum (the Tyrrhenian Sea). The names of the dark areas were those of bodies of water, and those of the bright areas, of lands. The brightest of all was Hellas—Greece itself. Eden and Elysium were also there. However, in a number of cases the bright areas bore names of deserts. Thus Libya and Arabia found their way onto his map, and one might argue that the theory of planetary desertification later associated with Percival Lowell was already subliminally present. Schiaparelli warned that his nomenclature "should not be allowed to prejudice the observations,"[42] hoping only that names whose sounds awakened such pleasant memories would be easily retained. Dugald Stewart, in a passage chosen by Beer and Mädler to introduce their section on lunar topography, had similarly argued that "phenomena should always be described by names that involve no theory as to their causes."[43]

The fact of the matter is that Schiaparelli—no doubt quite without any malice aforethought—had effectively refashioned Mars with a whole new set of romantic and wistful names, whose evocative power, in spite of his stated cautions, was not to be lost on the human capacity to yearn after lost paradises or to conjure up nostalgic visions. What Percival Lowell once said is no doubt very true: "Naming a thing is man's nearest approach to creating it."[44] In a sense, Schiaparelli's names created a new Mars, or at least a new way of looking at the old one. It is indeed interesting, and no doubt psychologically significant, that with the introduction of this new map of Mars with names so apt to appeal to human emotions, the planet began to gather around itself in succeeding years a considerable mythology of its own. Moreover, it is fitting that this map, which in its nomenclature put Mars into the realm of mythical rather than factual places, should also have been the first to include the strange "canals," which were to have such an important role in the planet's subsequent mythification.

"ALL THE VAST extent of the continents," Schiaparelli later wrote, "is furrowed upon every side by a network of numerous lines or fine stripes of a more or less pronounced dark color, whose aspect is very variable. These traverse the planet for long distances in regular lines, that do not at all resemble the winding courses of our streams."[45] So the "canals" later appeared to Schiaparelli in what became their most celebrated guise; but on first blush these curious configurations of the planet's surface looked neither fine nor, indeed, particularly regular, as is evident from his 1877 map (Fig. 7.4).

The first "canal" Schiaparelli detected would seem to have been the Cyclops on September 15, followed by the Ambrosia a week later and the Phison and the Ganges on October 4.[46] At this time Schiaparelli was using a magnification of 322× on the 8.6-inch Merz refractor. The canals early showed their predisposition for appearing only one or two at a time, rather than as a whole network, and their peculiar tendency to show no regard for distance. In the words of Percival Lowell, "Distance . . . is not, with the canals, the great obliterator."[47] Schiaparelli's own notes are well worth quoting.[48] On October 4, when the planet's disk was 21″ of arc across, he recorded that in the ochre region between the Pearl-Bearing Gulf and the Bay of the Dawn no "canals" were visible except for the Ganges, even though there were moments of perfect definition. "Of Indus, Hydaspes, Jamuna, Hydraotes," he says, "was no trace to be seen," and "neither was there any trace of disconnected spots." The same area of the planet was unchanged when studied again in early November (when, by the way, Schiaparelli made one of the first observations of a Martian "lake," describing a "poorly bounded but seemingly definite spot" where the Ganges intersected another "canal," the Chrysorrhoas. (In light of his tendency to see the Martian features in geometrical forms, it is perhaps worth mentioning that this was later to appear to him as "trapezoidal" in shape). Four months later, on February 24–26, 1878, the Indus was "easily visible" in the region that had hitherto been blank, despite the fact that by then the disk had shrunk to only 5".7 of arc.

Any features recorded so far from opposition, on a disk 5".7 across, should, if real, have been glaringly conspicuous at opposition itself. This point was later made tellingly by Schiaparelli's countryman Vincenzo Cerulli, who wrote of his own observations:

> From July, 1896 to February, 1897 [the lines] did not appear either to increase or decrease in their visibility. Several canals were already sufficiently easy in July, when the disk was only 7"; in December, when the diameter had increased to 17", they ought to have been magnificent; however, they presented on the contrary the same appearance as in July.[49]

There were other peculiarities about the "canals." As Schiaparelli continued to draw them they became ever straighter, yet if they were actually furrows following the shortest distance between two points on the globe of Mars, they failed to obey the laws of perspective. The fact was first pointed out by E. Walter Maunder in 1894,[50] and it was later seconded by E. M. Antoniadi, who wrote sarcastically:

> Behold, then, the arc of the great circle of the sphere at the central meridian, which is transfigured into an arc of a small circle at 35° meridian in order that each segment might remain perfectly rectilinear to the discoverer of the Martian canals on Earth. The reason of the scholarly world was never more gravely insulted than by these fables of the rectilinear canals of Mars, which rebel against the laws of nature; and the Martians, in the signals which they send us, truly put forth an ingenuity that surpasses our understanding."[51]

Percival Lowell later made a point of plotting his canals on small globes, perhaps in part to protect his representations from such criticisms. "Bearing upon Mars," he wrote to his secretary, Miss Wrexie Louise Leonard, "on the rectilinear appearance of the canals, is the clipping enclosed. . . . It explains why the lines do not appear curved as they should. The

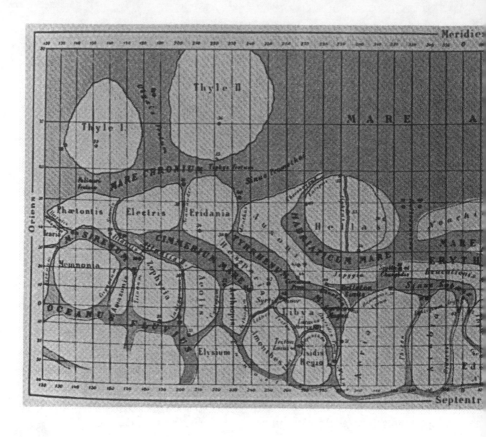

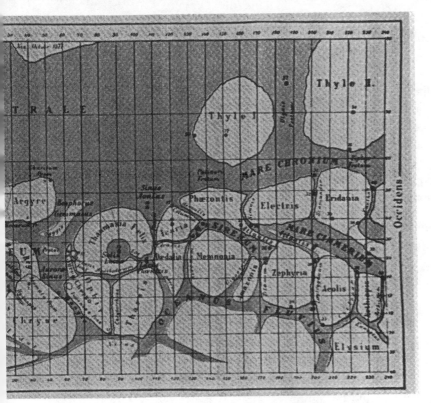

*Figure 7.4*
*Schiaparelli's 1877 map of*
*Mars, which introduced a new*
*nomenclature for the planet and*
*the famous—or infamous—*
*Martian canals.*

eye rectifies them."[52] Here, at least, as we shall see, he was on solid ground.

Schiaparelli's observations had yet to be critically analyzed. He himself was quick to point out that he had not been the first to see the new class of features, finding suggestions of them in the drawings of Dawes and others. Nor was the name *canale* itself his responsibility, that having been introduced into the Martian nomenclature by Secchi years earlier. But certainly no one else had shown them in anything approaching the same profusion. With Schiaparelli they became a dominant motif of the planet and made—whatever their implications for the question of life on the planet—the "analogy to the Earth," which had been assumed ever since Herschel's if not indeed Huygens's day, considerably less precise. Webb wrote of the appearance the Earth might have as viewed from Mars:

> There is every reason to believe that our surface would then appear mapped out by a distinct separation into oceans and continents, the fluid appearing darker than the solid masses, and preserving their bluish-green tinge but little affected by distance. . . . The general hue of the land would be lighter; and at a distance where its variegated patches would be separately undistinguishable, the result would be a grey resulting from the mixture of many tints, except where tracts such as the great deserts or prairies might subtend a sufficient angle to preserve their natural hue, or where extensive forests might rival seas in depth of tone.[53]

Though Webb admitted that with respect to the main features Mars presented "a very satisfactory agreement," this seemed to be abandoned when one came to the minute details. Martian research, so long dominated by the "analogy to the Earth" and moving confidently forward by finding at each successive stage of discovery more evidence of the supposed correspondence, had clearly entered, with Schiaparelli, upon a new phase.

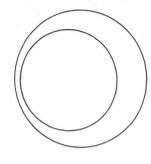

CHAPTER 8

# The Art of Observing

*The artist, clearly, can render only what his tool and his medium are capable of rendering. His technique restricts his freedom of choice. The features and relationships the pencil picks out will differ from those the brush can indicate. Sitting in front of his motif, pencil in hand, the artist will, therefore, look out for those aspects which can be rendered in lines—as we say in pardonable abbreviation, he will tend to see his motif in terms of lines, while, brush in hand, he sees it in terms of masses.*
    —E. H. Gombrich, Art and Illusion

EVEN AT the best sites, perfectly serene air is not attainable, and the passage of eddies in front of the telescope intermittently chops up the image and blurs the critical details. The duration of the intervals between interruptions is typically such, in fact, as to make the limits of the perceptual system determine how much, and how accurately, the information can be extracted. The eye, while quick, does not record instantaneously. Nevertheless, it does do better than the photographic plate, and this was true to a far greater extent in the late nineteenth century, when photographic processes were painfully slow. While photography soon proved its usefulness in recording faint stars and nebulas, it could not compete with the eye when it came to the planets.

One of the pioneers of celestial photography was Dr. Henry Draper, who obtained the first photograph of the Orion Nebula on September 30, 1880, at his private observatory in    *95*

New York City. Draper was a gifted and many-sided man.[1] A physician by profession, he set up his own physical laboratory and a splendid observatory equipped with an 11-inch Clark refractor and a 28-inch reflector Draper himself built. His father, John William Draper, had taken the first daguerreotype of the Moon in 1840, but the usefulness of photography in astronomy had been greatly limited by the insensitivity of the primitive wet-process plates.

Only a few men of advanced views were able to appreciate at the time the important role that photography would come to play in astronomical research. One of these was George Phillips Bond, who on the death of his father, William Cranch Bond, became director of the Harvard College Observatory. In the 1850s he succeeded in registering the images of Mizar and Alcor, the famous naked-eye double, on wet collodion plates with the 15-inch Merz refractor. At the time, he noted: "There is nothing . . . extravagant in predicting the future application of photography on a most magnificent scale."[2] However, Mizar and Alcor are fairly bright stars, and in Bond's own work on the Orion Nebula, carried out between 1859 and 1863, he had no choice but to make use of the old visual methods.

Asaph Hall remembered in later years, "how cold my feet were when [Bond] was making his winter observations on Orion":

> I sat in the small alcove of the great dome behind a black curtain, and noted on the chronometer the transits of stars when Professor Bond called them out, and wrote down also the readings for declination. . . . Sometimes I was called to the telescope to examine a very faint star, or some configuration of the Nebula. Prof. Bond had one of the keenest eyes I have ever met with. His work on this great nebula forms an epoch in its history.[3]

It would be difficult to estimate the number of hours that Bond spent preparing his final drawing of the nebula (Fig. 8.1). "He made scores of drawings, in white on black, and the reverse, in colors, etc.," in preparation for it, wrote Bond's cousin and later the first director of the Lick Observatory, Edward S. Holden, in his *Monograph of the Central Parts of the*

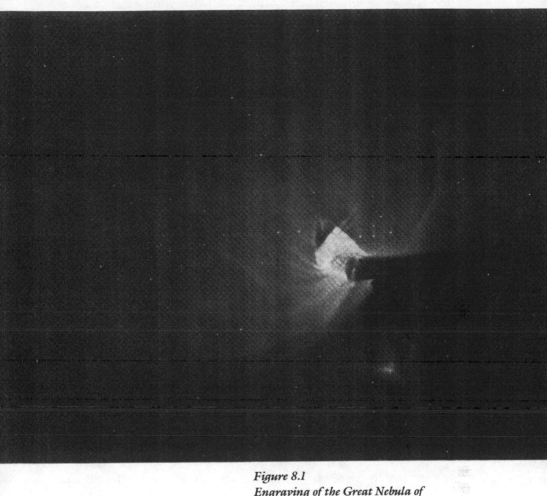

*Figure 8.1*
*Engraving of the Great Nebula of*
*Orion by J. W. Watts, based on*
*visual observations made between*
*1858 and 1863 by George Phillips*
*Bond with the 15-inch Merz*
*refractor of the Harvard College*
*Observatory.*

*Nebula of Orion.* Each of these, Holden continues, "was revised and re-revised many times. The revision of the . . . plate lasted many months, and I have myself examined from fifteen to twenty 'final' revisions of the plate. Color, form, and relative brilliancy were all successively and exhaustively criticized." The result of Bond's pains was what Holden called "the most accurate drawing that has been made, even as a map, and as a picture it is decidedly the best representation of a single celestial object which we have by the old methods."[4]

Yet for all the fastidious care bestowed on the drawing and Bond's exceptional skill as an observer and artist, a 137-minute exposure on a gelatino-bromide dry plate by Dr. Draper on March 14, 1882, showed the nebula, in Holden's opinion, "for nearly every purpose incomparably better than the other" (*Nebula of Orion,* p. 227). The plate could build up over hours a cumulative record of the light of faint stars or the outlines of nebulous forms beyond the reach of the eye's nearly instantaneous grasp—an advantage that was to increase tremendously in succeeding years with the introduction of more sensitive films, making visual methods, for such purposes, totally obsolete.

Yet in planetary observation the eye retained its supremacy over the plate. The slowness of the latter, no obstacle to the recording of faint objects such as stars and nebulas because of its ability to build up an image cumulatively, quite indisposed it to the recording of planetary surface details. The plate had to have time to work, yet in that time the details of a planetary surface were certain to be blurred hopelessly by the restless medium of the air. The eye, while lacking the plate's enormous depth, was in every respect its master when it came to agility. The atmosphere, heated unevenly from point to point, is here condensed, there rarefied. It is in constant motion, and this is experienced on a large scale as wind and on a small scale as turbulence and eddying. The denser the air, the more strongly it refracts a light ray passing through it, so that as regions of condensed and rarefied air pass successively in front of the telescope, the light ray is thrown hither and thither. This is the process that produces twinkling in the image of a star, wavering or boiling in the image of a planet.

The image of the planet—while generally more or less jumbled owing to these atmospheric disturbances—does nevertheless become sharp for an instant every now and then, and in these instants the finer details may stand out with startling abruptness. What the observer is faced with is similar to what would be experienced by someone watching a motion picture in which the camera is out of focus except for an occasional sharp frame thrown in at random.[5]

The eye is the optimal instrument for detecting what is so fleetingly revealed. Indeed, because it is sensitive not to absolute levels of intensity but to contrasts, there has to be relative motion between the object viewed and the eye for seeing to take place at all. Even if the object itself is at rest, the eye is scanning constantly, and despite every effort to hold the gaze fixed, the eye continues to make small unconscious excursions. Indeed, if the image *were* somehow fixed on the retina (say, by means of small mirrors attached to the eye), it would fade out in a matter of seconds.[6] Thus the eye is quite different from the photographic plate. A long exposure does not build up an image; quite the reverse—it causes it to disappear. Perfect steadiness is fatal to it. At the same time, because the eye is designed to be scanning continually, the critical duration for visual acuity is less than a tenth of a second (the time needed for the retinal image to scan through an angle less than that of the separation between adjacent single cones). For longer exposures acuity improves, but only up to exposures of about 0.2 seconds or so (for black test lines). After this there is no further improvement in visual acuity.[7]

The interludes of sharp visualization of planetary detail allowed by the atmosphere tend, as a general rule, to be roughly of this order of duration. Evidently for very short exposures of up to 0.2 seconds or so there will be a gradient of image quality—the eye, though fast, is not quite instantaneous. The duration of the "revelation peeps," as Percival Lowell once called them, will thus have a great deal to do with defining how much of the available information the eye-brain system will be able to retrieve, a point we shall return to later.

Of course, perception obviously involves more than the mere reception of the data. It involves classifying the image in terms of some characteristic feature that allows its identity to be grasped. If the object is a familiar one, well and good. If not, the recognition must be approximated through some intermediate category.

Experiments with perceptions during short exposures have demonstrated that success in recognizing what is then presented depends on whether one can find an appropriate schema for classifying it. If one can, then it can be filed in terms of this symbol, using a summary notation or shorthand. A large part of perception involves, then, an encoding process, and it should hardly surprise us that the simplest schemata—such things as lines, geometrical figures, and the like—often mediate this process. Such schematic forms, indeed, serve to catch *something* of what is revealed, though not necessarily doing it full justice. Given the nature of planetary perceptions, it should hardly be surprising that schematic forms entered into Schiaparelli's perceptions of Mars in the geometrical forms he gave to delicate traceries of Martian "continents."

Then, too, one might wonder what role idiosyncrasies of vision or temperament may have played in further exaggerating the tendency. We shall have more to say about Schiaparelli's vision later. As for his temperament, there was his training as an engineer—more particularly, as a hydraulic engineer. Despite his insistence on the tendency of nomenclature to prejudice observations, he himself later came around to interpreting the Martian features consistently with their names—describing them as making up a "hydrographic system"—so that the nomenclature, innocent though the basis on which it was introduced may have been, seems by then to have taken on a life of its own. The tail, as it were, started wagging the dog. What role did Schiaparelli's nomenclature play in unconsciously fleshing out the marginal material on which his perceptions were based?

Finally, and perhaps most important, there was the undeniable hardness and sharpness of the features as Schiaparelli represented them, which made them look so unnatural. He had approached the planet in 1877 with the object of "triangulating" the positions of the surface features, of finding out their "plan." The precision he used in his measurements and the care he took in reducing them were, of course, admirable in their way, but one wonders whether this preoccupation with precision, with fixing the indefinite definitely, may not have been misapplied when it came to the recording of these highly uncertain details. In 1899, when he had virtually given up observing the planet himself, he encouraged the younger generation of astronomers who were taking over for him to make their work "precise measures! the thing most necessary and at the same time the most difficult."[8] It was his own obsession. But did he, perhaps, bestow upon the fugitive objects as well as the definite ones a wished-for sharpness and precision that they lacked in themselves, making the whole planet over, in fact, in the terms imposed by the idiosyncratic tendency broadly spoken of as style?

Certainly when he said on a number of occasions that his *canali* appeared to be "of absolute geometrical precision . . . as if drawn by rule and compass,"[9] one can only say that absolute precision was his great theme and that the rule and compass were the instruments in which he was, by training, himself most expert. In astronomy, as in any other study, the saying of Dr. Johnson no doubt rings true: "No man forgets his original trade."[10] Or as Francis Bacon said with his usual acumen:

> Men have used to infect their meditations, opinions, and doctrines, with some conceits which they have most ad-mired, or some sciences which they have most applied: and given all things else a tincture according to them, utterly untrue and improper. So hath Plato intermingled his philosophy with theology, and Aristotle with logic, and the second school of Plato, Proclus and the rest, with the mathematics. For these were the arts which had a kind of primogeniture with them severally.[11]

Just as Schiaparelli's nomenclature unmistakably revealed the classicist, his system of "canals" revealed the hydraulic engineer with an obsession for precision. The best observations are the most impersonal ones, but these observations of Mars clearly had Schiaparelli's personality stamped all over them. What appeared in them was evidently as much an artifact of style as an insight into the structure of a planet.

IN SUBSEQUENT YEARS, after the canals (for such we shall henceforth call them, without apologizing for them with quotation marks) first came to Schiaparelli's attention, they became ever sharper and more regular. One sees this tendency already at the next opposition, that of 1879. His drawings look almost as if he had suddenly developed an enthusiasm for imitating the geometric age of Greek art (and indeed this was,

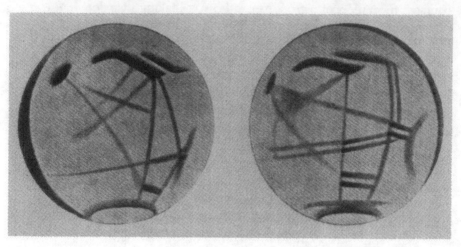

*Figure 8.2*
*Representative drawings of Mars*
*by Schiaparelli with the 8.6-inch*
*Merz refractor at the Brera*
*Observatory, Milan: December*
*22, 1883 (left), and February 27,*
*1884 (right). Note the bizarre*
*appearance of the surface features*
*compared to those shown in his*
*1877 map (Fig. 7.4).*

representation). In addition, one of the canals had become
double, the first instance of what came to be known as a gem-
ination. By 1881 most of the canals depicted on Schiaparelli's
map had taken this form—like Noah's animals, they now
came two by two. The appearance of the surface had become
decidedly unnatural, and there was little alternative but to
think that Schiaparelli was the victim of some peculiar illusion, that the doubles were duplistic indeed, equivocating with
him like Macbeth's witches, who "palter with us in a double
sense." If they were real, there could be scarcely any explanation for them in nature. Though other astronomers were soon
to confirm in abundance the canals themselves, so that in almost every map published between 1878 and 1903 they appeared as a conspicuous feature, the geminations—though
seen in a few special cases by all of the more prolific canalists—were never, one might say, duplicated to quite the same
extent as on Schiaparelli's maps. One might suppose, therefore, that the detection of the geminations was a matter of
some difficulty. Yet according to Schiaparelli, "In the Milan
instrument the greater number of the canals and of their pairs
were observed with comparative ease whenever the air was
still, and only a few cases required a special effort on the part
of the observer." He even suspected that the single canals that
remained were actually double, but his telescope could not
resolve them, and they "showed in that place a large, broad,
and somewhat confused stripe." Finally, he noted that "some
of the pairs show so great a regularity that one would say that
they were a system of parallel lines drawn by rule and compass," though he added that "perhaps . . . this regularity will
not resist the use of a high magnifying power."[12]

SCHIAPARELLI WROTE of his canals to the English astron-
omer-artist Nathaniel Green in 1879 that "it is [as] impossible
to doubt their existence as that of the Rhine on the surface of
the Earth."[13] This was nothing if not definite. Yet Green himself had carefully observed Mars in 1877 with his 13-inch reflector from the island of Madeira (in the Atlantic off the coast
of Morocco, where the skies were certainly as splendid as those
of Italy; see Fig. 8.3), and in the course of compiling there a

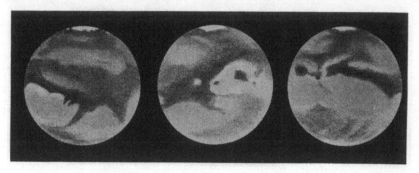

*Figure 8.3*
*Drawings of Mars by Nathaniel E.*
*Green with a 13-inch reflector at*
*Madeira in 1877.*

justly celebrated map of the planet (Fig. 8.4), he had noted nothing unusual.[14] Green observed the planet again in 1879, but from St. John's Wood, London, rather than from Madeira, and he complained that unfortunately "the definition afforded by the St. John's Wood atmosphere has barely sufficed to identify the details of the Madeira drawings," much less any new ones. However, he added that "faint and diffused tones may be seen in places where Professor Schiaparelli states that new canals appeared during this opposition."[15]

At the same opposition the Irish observer Charles E. Burton, also using a modest reflector, was somewhat more successful, managing to make out traces of forms similar to the canals of Schiaparelli.[16] Green, in comparing Burton's drawings to those of the Milan astronomer, noted, however, that the two observers did not agree on the positions in which they showed the canals, and concluded: "It is hardly safe to regard them as belonging to the permanent markings." Instead he urged his own preferred explanation of them. "It is possible," he wrote, "that some of these lines may be the boundaries of faint tones of shade."[17] Two years later some of the canals were made out by E. Walter Maunder, who later became one of the leading figures in the cause célèbre. Summarizing his own observations at the opposition of 1881, he concurred with Green on the main point:

During the recent opposition I had a pretty fair view of one side of Mars; and there certainly seemed to be a number of spider-like markings radiating from the center in different directions, which coincided roughly with some of the canals shown by Prof. Schiaparelli in his 1879 observations. But at the same time I ought to say that my drawings of the same district on different nights do not perfectly agree. The canals, if you adopt the word, which I have seen on one night have not been always detected in the same place on another night, and fresh canals in different directions have appeared to take their place. . . . [W]here I have represented shaded districts, [Schiaparelli] has drawn hard lines corresponding with the borders of those districts; so that where he has given a number of parallel and interlacing lines, I should myself have rather shown faint shaded districts between those lines.[18]

Thus, though there seemed to be some basis in reality for what Schiaparelli had seen, the consensus was that he had not depicted the Martian features in their true form. Green was later to make this criticism quite explicit: Schiaparelli, he insisted, and others who drew the Martian surface in the same way, "have not *drawn* what they have *seen,* or, in other words, have turned soft and indefinite pieces of shading into clear, sharp lines."[19]

GREEN'S CRITICISM was perhaps only to be expected of one who was a professional artist as well as an astronomer— and an artist whose skill was sufficiently recognized that he could be called upon for instruction in painting by no less a personage than Queen Victoria herself. His own 1877 map of Mars was incontestably more *artistic* than Schiaparelli's, being very carefully shaded and resembling a "picture" rather than a "plan," as Rev. T. W. Webb happily said of the rival depictions.[20] Indeed, Webb's comments are worth quoting in some detail. Speaking of Green and Schiaparelli, he noted that "each did his best; each was far in advance of the other observers of the season." However,

> at first sight there is more apparent difference in their results than might have been expected. It is not surprising that in the case of minute details each should have caught something peculiarly his own; but there is a general want of resemblance not easily explained, till, on careful comparison, we find that much may be due to the different mode of viewing the same objects, to the different training of the observers, and to the different principles on which the delineation was undertaken. Green, an accomplished master of form and colour, has given a portraiture, the resemblance of which as a whole, commends itself to every eye familiar with the original. The Italian professor, on the other hand, inconvenienced by colour-blindness, but of micrometric vision, commenced by actual measurement of sixty-two fundamental points, and carrying on his work with most commendable pertinacity, has plotted a sharply-outlined chart, which, whatever may be its fidelity, no one would at first imagine to be intended as a representation of Mars. His style is as unpleasantly conventional as that of Green indicates the pencil of

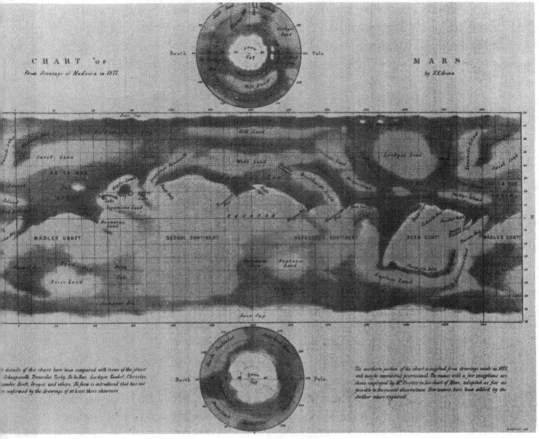

*Figure 8.4*
**Nathaniel Green's map of Mars,
1877.**

an artist; the one has produced a picture, the other a plan. The discordance arising from such opposite modes of treatment would naturally be less real than apparent; still, a good deal remains that is not easy to harmonise.[21]

Proctor, in a similar vein, went so far as to comment that "no one who has ever seen Mars through a good telescope can accept the hard and unnatural configurations depicted by Schiaparelli."[22] If Proctor may have been biased, considering that his system of nomenclature (followed in the main by Green) had been unceremoniously discarded by the Italian observer, other astronomers without his possible ulterior motive concurred in judging Green superior. The American astronomer Charles Augustus Young would write of his impressions of Mars with the 23-inch refractor at Princeton in 1892, when the planet was again presented under conditions similar to those of 1877: "The appearance of the planet in general corresponded much more closely to the drawings of Green, made at Madeira in 1877, than to any others with which I am acquainted."[23]

The same year James Keeler of the Lick Observatory compared his own drawings of Mars with a 6-inch globe based on Schiaparelli's work in an attempt to determine whether the differences could be better explained as a result of real changes in the surface or as mere artifacts of drawing style. He emphasized the "importance for such comparisons in representing as truthfully as possible not only the *forms* of the different markings but also the precise intensity of their shades and their character, whether sharp or diffused," pointing out that "mere diagrams of the surface were exceedingly misleading . . . as it was an especial characteristic of Mars that many of its markings had no sharply defined outline whatever, and though a few were distinctly marked, the majority were diffused."[24] In other words, they were quite unlike the appearance given to them by Schiaparelli.

When Keeler's paper was read to the Royal Astronomical Society by E. Walter Maunder at its meeting that year, Green rose to comment on it and commended Keeler's own drawings as possessing "the valuable quality of representing the correct forms, but also the proportional depth of shade of the various markings, together with that softening of the edges that is so sadly neglected by some draughtsmen."[25] Clearly, Schiaparelli was the chief offender, and while mentioning no names, Green unmistakably had the Italian astronomer in mind as the most illustrious of those practicing the sort of deceptive drawing to which he objected:

> I feel assured that the hard, straight-edged, diagram-like shapes, so frequently shown, do not exist, nor indeed is there anything like them on the surface of Mars. On the contrary, the planet is covered with the most delicate mottling of shapes and streaks of shade, and it requires a long training of the eye to see and follow the extent of these shades, and a considerable training of the hand to imitate them when seen. These faint shades fill up the spaces between the usually definite dark marks just where some observers place the canal systems, and it is these faint forms of shade that are totally wanting in the hard drawings to which I allude.[26]

Schiaparelli himself indicated on at least one occasion that his *published* drawings, at any rate, did not necessarily satisfy himself as to the true appearance of the Martian surface markings. When the American astronomer E. E. Barnard visited him in Milan in 1893 and had the opportunity to examine his observing notebooks at firsthand, he wondered at the difference between the sketches and the published drawings and asked: "Is it an accident of the reproduction that they are so heavy and dark in the engraving?" Schiaparelli admitted that the reproductions of his drawings could mislead but placed the blame on his engravers: "I cannot find artists who reproduce them well."[27]

Whether the fault was more with Schiaparelli or his engravers, certainly the Italian astronomer had no right to criticize Percival Lowell's later maps for being "too schematic," or to complain that the canals had been shown by Lowell without sufficient attention to their differences in intensity or breadth.[28] Precisely the same problems could be cited with respect to his own drawings. They too were schematic, and though the features had been represented with a micrometric exactness, they tended to be shown, in Webb's words, in an "unpleasantly conventional" style. Lowell's drawings, in fact, point toward the next development. The observers of Mars who had learned their way around the planet's surface by direct telescopic work before Schiaparelli's charts of Mars were published—like Green, Webb, and Proctor—had no difficulty recognizing the latter as stylized representations, even gross caricatures. But as those maps became widely disseminated and discussed, a number of observers—Lowell being the most significant—became acquainted with the planet only, as it were, through Schiaparelli's eyes. So it is no wonder that Lowell's drawings distorted the Martian features in just the way that the Italian observer's had.

Indeed, canals, represented more or less as on Schiaparelli's charts, were all the rage in the 1890s. What role did suggestion play in this? To what extent was the Martian canal "furor" of the 1890s a kind of scientific equivalent of mass hysteria? The late Victorian age was, of course, one in which scientific circles were abuzz with talk of hypnosis and suggestion. Pierre Janet, with Freud a founder of the psychology of the unconscious, had held suggestion to be a relatively rare phenomenon found only in hysterical individuals. But others disagreed, maintaining that the suggestibility of hysterical individuals did not differ in kind, but only in degree, from that mental suscep-tibility that is common to all of us, which makes us "yield assent to outward suggestion, affirm what we strongly conceive, and act in accordance with what we are led to expect."[29]

As for Schiaparelli himself, he continued to publish his own subsequent observations of Mars in a succession of elaborately detailed memoirs. During his lifetime his reputation was enormous, and even today one must admire the diligence with which he recorded the most minute observations. What he recorded, however, sometimes seems more a testimony to the power of his sight than to the power of his insight.

Camille Flammarion in France and Percival Lowell in the United States hailed him as the "Columbus of Mars." This is hardly surprising, in light of their own views about the planet. Yet even the conservative W. W. Campbell, one of the severest critics of Lowell's later observations and theories, began his review of Lowell's 1895 book *Mars* by eulogizing Schiaparelli: "The reviewer of a work on organic evolution," he wrote, "would find it difficult to avoid mentioning Darwin. Schiaparelli holds a similar place in the literature of Mars. An intelligent criticism of any recent book on Mars must consist of a review of Schiaparelli's observations and ideas." And he continued,

> Schiaparelli's work extends continuously from 1877 on. It is impossible to do justice to his labors in this article. He extended our knowledge of the planet enormously in nearly every line—in reference to the polar caps, the so-called seas and continents, but especially in reference to the so-called canals, their appearance and disappearance, their doubling, etc. His entire work bears the impress of a scientific spirit *par excellence*. His observations cover the period 1877–92, but his technical results are comprised in a few papers, and a dozen octavo pages suffice for a masterly popular exposition of his general results. His brief paper contains at least the suggestion of all the theories recently exploited by popular writers, though he was not concerned with establishing a theory, but rather with ascertaining the facts.[30]

Campbell's view of Schiaparelli as one who was only concerned with "ascertaining the facts" was shared by E. S. Holden, also of the Lick Observatory, who called Schiaparelli one who "has been very chary of hypotheses. . . . Most of his writings have been concerned with the pure results of observations, and he has scrupulously refrained from generalizations."[31] Unfortunately, as we have seen again and again in our discussion of the visual observers, observation is not to be so neatly sundered from theory as might be supposed. The dichotomy is not between observation and theory. To the contrary, sight becomes insightful only when it is informed by some "object-hypothesis" about the true nature of what one is observing. The observers who recorded Saturn "under a succession of . . . strange and marvelous forms" may have been, like Schiaparelli, "very chary of hypotheses." Yet it was only when Huygens invented the hypothesis of the ring that the observations finally became intelligible.

Similarly, having the courage to draw conclusions from the observations—even the most extreme conclusions—may be the most important step in eventually recognizing the truth. Any hypothesis, even if wrong, is generally better than none at all, because it can be subjected to stringent tests and, if wrong, disproved. Lowell, by coming forward with his theory of intelligent life on Mars, is often cast as the villain of the whole canal controversy. Yet certainly the theory he proposed was a logical enough conclusion if one accepted that there really were markings on Mars of the form that he recorded. Schiaparelli gave the impression of avoiding hypotheses, but in fact, to the end of his life he defended the all-important one: that the canals and their geminations were real, and that the whole "system" (a word whose applicability to these features even he acknowledged in later years) actually existed. Though he hesitated to draw conclusions, he, no less than Lowell, believed in the literalness of what his observations were revealing.

Antoniadi later said that though he was "innured for years to the fleeting visibility of straight lines on Mars with the ordinary appliances he was using," he had begun to grow skeptical of these observations because of the "difficulty of reconciling the 'canal' phenomena with logic."[32] Thus, only when he tried to draw logical conclusions from the observations did he begin to grow skeptical of the canals. Schiaparelli, with his obsession with observational precision, exaggerated the accuracy of which the human eye working under such difficult conditions is capable. He seemed to be humble in taking the lowly way of observation and in leaving the theorizing to others. In fact, not only did he establish a false dichotomy between observation and theory, he showed no less hubris on the observational side than others did on the theoretical side. Indeed, we find that in doing so he was, ironically, largely responsible for the theoretical excesses as well— for as Antoniadi again put it, it was only by "a disregard of the dangers of pressing too closely the evidence of the senses that some observers framed those startling theories about the planet."[33] In retrospect we can say that visual observers in general spent too much time bothering with markings at the uncertain threshold of perception.

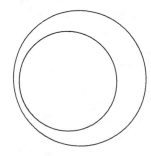

CHAPTER 9

# Of Aperture and Atmosphere

*Immense telescopes are only about as useful as the enormous spectacles which are suspended over the doors of opticians!*
— Dr. William Kitchiner

NEITHER ASAPH HALL nor William Harkness at the U.S. Naval Observatory were able to see any of the Schiaparellian canals in 1877, despite using a telescope with nearly three times the aperture of Schiaparelli's. It is true that their observations of the surface of Mars did not extend past October 18, and many of the canals did not appear even to Schiaparelli until later in the observing season. Yet it was still strange that the chart published by the Washington observers, made using the largest refractor in the world, not only failed to show the mass of complicated details depicted by Schiaparelli but, as William F. Denning pointed out, it showed no more than could be seen with a good 4-inch refractor.[1]

Eyesight no doubt had something to do with this. Schiaparelli's eyes were regarded as exceptionally keen. Perhaps the difference in what was recorded at Milan and at Washington could be attributed to a mere difference in visual acuity. Cases were reported of persons with apparently miraculous vision. There was, for instance, I. W. Ward, of Belfast, who claimed to be able to see two of the satellites of Uranus and one of Neptune with only a 4.3-inch glass—a remarkable feat, if true. Then there was Adolphe de Boë, the nephew of Baron von Ertburn, a Belgian observer, who was said to have made out two new companion stars of Polaris with a 6.3-inch glass—

objects which had eluded, and continued to elude, the observers at Washington. Admittedly, these achievements, or alleged achievements, involved the detection of faint stars. But why should planetary observation be any different? Indeed, Captain William Noble, in an article published in 1877 in which he pointed out that differences in sensitivity to colors may account for discrepancies in stellar brightness estimates between different observers, noted that both Ward and de Boë were evidently skilled in the observation of planetary details as well. Noble concluded that in certain persons there is "an acuteness of vision which seems to have no assignable reference to colour (as such) whatever; but possibly to have its origin in an exceedingly accurate adjustment of the various parts of the eye, *inter se,* and in an abnormal sensitiveness of the optic nerve."[2] He cited as an instance of this Rev. Dawes, who, as he recalled, had been able to detect the known companion of Polaris

> with an object-glass of only 1.6 inches in aperture. His delicate perception of planetary markings, &c., is also equally notorious. I mention this here . . . because a supposition exists that a person whose optical idiosyncrasy renders them [*sic*] peculiarly perceptive of faint planetary markings, are among those who fail most conspicuously to pick up very minute points of light.[3]

The supposition that Dawes appeared to refute, in Noble's eyes at least, would be raised again presently. But as far as the claims of Ward and de Boë of having "miraculous vision" were concerned, these soon met with the following sharp criticism from Herbert Sadler, who wrote in response to Noble's article:

> It has happened before now, unfortunately, that supposed companions to stars have been seen and measured. . . . If such an experienced observer as Otto Struve could err, if Admiral Smyth could enter, not one, nor two, but many stars in the cycle as having companions which do not really exist, I think I may be pardoned a doubt if M. de Boë's companions are not due either to his imagination, or his telescope, or both combined.[4]

And for good measure, he added, referring to the claim that
de Boë was able to see the stars "without rays" (that is, apparently, without aberrations of any kind):

> Mr. Ward, I believe, also sees stars without rays, and I
> have heard of other cases. If this extraordinary child really
> can count the separate larks in the flocks before ordinary
> mortals can see a trace of the flocks themselves . . . I can
> only suppose that by some means or other he has suc-
> ceeded in getting hold of a pot or so of the miraculous
> ointment the dervish used.[5]

Though such criticism of these rather exaggerated claims
may have been just, the possibility that there may now and
again be an individual blessed with exceptional eyesight—with
"an exceedingly accurate adjustment of the various parts of
the eye, *inter se*," as Noble put it—was certainly reasonable.
Schiaparelli was the leading "seer" among professional astronomers. But who was to say what still more lynx-eyed persons
might be found among the legions of amateur astronomers.
Despite the modest apertures they used, they might nevertheless stand a chance of defeating those armed with far larger
telescopes. It seemed almost self-evident, in the light of what
some—Dawes, for instance—had done with modest means,
that arbitrary limits could not be assigned to a given instrument in the hands of the right observer. To the contrary, the
powers of the telescope seemed to depend "in a very great
measure on the powers of the observer" himself.[6]

AT LEAST as important as the observer's eyesight was the
medium through which the observer had to view: the atmosphere of Earth. The kind of atmospheric disturbances
touched on briefly in the last chapter, though general across
the Earth's surface, are obviously worse in some places than in
others. The Italian skies were supposed to be relatively free of
them. The Washington telescope, on the other hand, was
handicapped by its unfortunate location in the lowlands along
the Potomac, which was almost as unfavorable a site as could
possibly be imagined, testifying to the almost complete indifference to the matter of where to station a large telescope that
prevailed even as late as the 1870s.

Hall's depressing experiences with the Washington telescope have already been recounted, and there is no need to repeat them here. Suffice it to say that the bad conditions he encountered during his quest for the Martian satellites were hardly exceptional: in 1887, according to the records of the observatory's night watchman, 168 nights were cloudy, and of the rest, 133 were rated poor, 43 fair, 19 good, and only 2 very good.[7] Eventually, in 1893, the telescope was moved to its present location on higher ground in northwest Washington.

As another case in point, consider the 25-inch refractor built by the Englishman Thomas Cooke for a wealthy amateur, R. S. Newall. For a brief time it was the world's largest, surpassing the 18½-inch Clark at Dearborn (only to be surpassed in turn by the 26-inch Clark at the U.S. Naval Observatory). It was set up in a dismal situation—at Newall's estate at Gateshead, across the Tyne from Newcastle in northern England. During fifteen years following the telescope's installation in 1869, Newall later admitted, he had only one night on which he could use the full aperture to advantage. He concluded: "Atmosphere has an immense deal to do with definition."[8]

As early as 1704, Sir Isaac Newton had written in the *Opticks*:

> If the Theory of making Telescopes could at length be fully brought into Practice, yet there would be certain Bounds beyond which Telescopes could not perform. For the Air through which we look upon the Stars, is in a perpetual Tremor; as may be seen by the tremulous Motion of Shadows cast from high Towers, and by the twinkling of the fix'd Stars. . . . Long Telescopes may cause Objects to appear brighter and larger than short ones do, but they cannot be so formed as to take away that confusion of the Rays which arises from the Tremors of the Atmosphere. The only Remedy is a most serene and quiet Air, such as may perhaps be found on the tops of the highest Mountains above the grosser Clouds.[9]

Newton, in grasping the importance of atmospheric conditions to telescopic seeing and in particular in recommending that telescopes ought to be placed in the "most serene and quiet Air," was certainly far ahead of his time; such considerations did not enter the minds of many of those who decided where the great telescopes of the nineteenth century ought to be placed. However, what Newton had seen by genius so long before was gradually confirmed by hard experience. The Newall and Washington telescopes were both refractors; but susceptible as *they* were to atmospheric disturbances, reflectors of the day, with their open tubes and vulnerable speculum-metal mirrors, were even more so.

William Lassell, who had built reflectors of 2- and 4-foot aperture and had set them up at Malta specifically for the purpose of obtaining more favorable conditions for observation, went on record as saying: "It is true that the defect of flexure [of large mirrors] may in some degree be eliminated, but that of atmosphere is quite unassailable. These circumstances will always make large telescopes *proportionately* less powerful than small ones."[10]

William Parsons, the earl of Rosse, whose 3- and 6-foot reflectors at Birr Castle were wonders of the age and first revealed the spiral form of some of the nebulas, found in his experience with these behemoths that

> a stream of heated air passing before the telescope, the agitation and hygrometric state of the atmosphere, and any difference of temperature between the speculum and the air in the tube, are all capable of injuring, or even destroying definition, though the speculum were absolutely perfect.
>
> The effect of these disturbances is, in reflectors, at least as the cube of their apertures; and hence there are few hours in the year when the 6-foot can display its full power.[11]

In the great Rosse telescope, according to G. Johnstone Stoney, who was in charge of it for a time, stars were bright but usually appeared "as balls of light like small peas, violently boiling in consequence of atmospheric disturbance."[12] He added, however, that "if the night is good, there will be moments now and then when the atmospheric disturbance will abruptly seem to cease for a fraction of a second, and the star is seen for an instant as the telescope really presents it." Sir George Biddell Airy remarked that "the orb of Jupiter produced an effect compared to that of the introduction of a

coach-lamp into the telescope."[13] Nevertheless, as W. F. Denning later pointed out, drawings of Jupiter made with the
Rosse reflectors showed no more details than would show up
in much smaller instruments. In fact, he said, a 6½-inch telescope would show them all "without any difficulty," and he
concluded that though

> it may sound imposing to describe . . . the advent of
> Jupiter into the field of a very large reflector as "like a
> coach-lamp," or the approach of Sirius near the field of a
> similar instrument as "like the dawn of morning," and its
> real ingress as accompanied "with all the splendour of the
> rising Sun" . . . I think there is something more attractive
> in the pale disk of a planet as it comes unannounced into
> the field of a small aperture and displays a still, sharply
> cut outline, enabling the most minute detail to come
> steadily out with all the exquisite clearness of an
> engraving.[14]

Denning, who lived at Bristol, was himself one of the most
skillful planetary observers of the day, though his specialty was
the study of meteors. He once remarked: "I never go out without expecting to see something of interest."[15] Evidently he was
seldom disappointed, for he discovered five comets and, late in
life, a nova (Nova Cygni of 1920).

Denning threw the gauntlet down to observers using large
telescopes, challenging them to show that they could outperform smaller telescopes when it came to planetary work. Referring to the huge expense of the large telescopes compared
with that of the modest equipment used by amateurs such as
himself, he wrote in 1885:

> Many people would consider that in any crucial tests the
> smaller instrument would be utterly snuffed out; but such
> an idea is entirely erroneous. What the minor telescope
> lacks in point of light it gains in definition. When the
> seeing is good in a large aperture it is superlative in a
> small one. . . . We naturally expect that very fine tele-
> scopes, upon which so much labour and expense have
> been lavished, should display far more detail than moder-
> ate apertures; but when we come to analyse the results it

is obvious such an anticipation is far from being realised. The glare of excessive light and the endless mouldings and flaring of the image can only have one effect in obliterating delicate markings.[16]

Denning had entered the fray in response to criticisms of his and other amateurs' observations of Jupiter that had seemed to establish the existence of rapid atmospheric changes on that planet. The criticisms came from a noted Jovian observer—and professional astronomer—G. W. Hough, who used the 18½-inch Dearborn refractor for his own observations. Failing to note the changes reported by the amateurs, he dismissed their reports out of hand as being due to "the poor quality of the images" in the small telescopes. Denning not only reaffirmed what he had seen but wondered how such changes could have been missed by Hough with such a large telescope, and he took advantage of the opportunity to underscore the dismal results that had generally been achieved with large telescopes in planetary astronomy. "Apertures of from 6 to 8 inches seem able to compete with the most powerful instruments ever constructed," he stated, and he gave as the reason that even when the images in the small aperture appeared sharp and still, "a very large aperture shows the rushing of vapours across the disc, and violent contortions of the image, which are the inevitable result." In support of his position Denning cited, among other things, the bland map of Mars drawn by Hall and Harkness, noting that not only Schiaparelli's drawings but also those of Dawes and Green, "though founded on views with much smaller instruments," showed far more numerous details.[17]

A number of observers responded to Denning's editorial. Asaph Hall was sympathetic and admitted: "There is, perhaps, too much skepticism on the part of those who are observing with large instruments with regard to what can be seen with small ones." He added, moreover, that based on his own observations of Saturn, "it is plain that my work will confirm Mr. Denning's criticism that the large telescope does not show enough detail."[18] With the exception of the white spot he had discovered on the equator of Saturn in December 1876, Hall recounted that he had seen no changes in eleven years of regular observation. Nathaniel Green noted that he had felt it necessary to equip his 18-inch reflector with a "convenient gradation of stops."[19] Here Denning could not resist a gibe:

> If a large diameter is useless without stops, wherein does its utility consist? Better at once adopt a smaller speculum and obviate the more troublesome manipulation of a large instrument. True there are very rare occasions when all the aperture may be utilized; but are they worth waiting for, and when they come do the results answer expectation?[20]

Though a number of astronomers who had used large telescopes thus confirmed Denning's view of their unsuitability for planetary work, a notable exception was the distinguished American astronomer Charles Augustus Young, who, though not a planetary specialist, had made a number of planetary observations with the 23-inch Clark refractor at Princeton, New Jersey. Though Young admitted that the "smaller beam of light incident on a smaller object-glass is less affected by atmospheric disturbances, and so, *quoad hoc,* with a given magnifying power the image is better defined," he believed that on the basis of his own experience the 23-inch was immensely superior to the observatory's 9-inch on one night in three.[21] He did not agree with Lord Rosse's cubic relation of maldefinition to aperture; that, in any case, had been derived from experience with large reflectors, and from his own work with refractors Young concluded that the relationship was more nearly as the square root, "if not a still more slowly growing function." Subsequently he and an assistant performed some experiments with diaphragms to see whether the full aperture would outperform the aperture stopped down to various sizes. Saturn was the test object and was examined with a magnification of 450×, first with the full aperture and then with the aperture capped to 9 inches. There was no noticeable difference, and at first Young felt "obliged to admit that we could see fairly with 9 inches about everything we have been able to make out with the whole aperture." But on removing the cap again, he said,

> This time there was a change. At our first observation the planet was only 25° high, and not fairly out of the haze.

Now . . . it was nearly 40° high, and although the air was still boiling, it had cleared the mists, and shone out splendidly. . . . A faint dark streak came out just on the equator. . . . The tropical belt showed, though not very clearly, a pronounced filamentary structure like that of the belts of Jupiter, and the polar cap, inside the Arctic circle belt, showed faintly its peculiar greenish hue.

On ring B a faint, ill-defined, concentric structure was indicated, a sort of "brush-marking," as if the ring were a painting made with a flat-brush, not by radial spokes, but by a circular sweep. There were no distinct stripes that the eye could seize upon, and yet the general impression was quite decided.[22]

Young's impression of the B ring could be compared to the remarkable revelations obtained by Charles W. Tuttle in October 1851 with the 15-inch refractor at Harvard,[23] yet as Young pointed out, this had been only a fair night.

Young, it is true, had still failed to pass Denning's most stringent test. Denning had pointed out that no new discoveries concerning planetary surface markings had been made with large telescopes. To the contrary, all the important discoveries—the new class of markings on Mars being a case in point—had been made with smaller apertures. But the supposed superiority of smaller instruments did not, at any rate, comport with Young's experience. On the contrary, he noted that "most commonly . . . when I have failed to see with the large instrument anything I supposed I saw with the smaller, it has turned out on examination that the larger instrument was right, and that imagination had constructed a story that was not true by building up faintly visible details and hazy suggestions furnished by the smaller lens."[24] A remarkable insight, yet Denning, at least, did not accept it, and he who had had the first word in the argument had the last word too. "The success of men like Dawes and Schiaparelli," he insisted,

who outstrip their contemporaries and with small instruments achieve phenomenal results, is to be ascribed to keenness of vision and to that natural avidity and pertinacity uniformly characteristic of the best observers. This may explain the fact why the eminent director of the

Milan Observatory sees more detail on Mars, with a refractor of only 8-inches, than is revealed to ordinary eyes with the finest instruments. It is a great fallacy to conclude that an object shown in a small telescope is a false appearance because it cannot be distinguished in a large one. . . . When the "canals" on Mars were first announced the idea was ridiculed in certain quarters, but the tendency of subsequent researches in the same direction has gone to substantiate these features. Step by step the great instruments appear to be grasping the complicated structure of delicate markings so obvious to Schiaparelli with his relatively diminutive glass.[25]

To this Young could only answer that what Denning said was probably true. "It is very doubtful," he admitted, "whether my eye, even with our great telescope, could see everything that other 'eagle-eyed' observers have made out with smaller instruments." But he did not doubt that such keen-sighted observers would see still more on the planets with a large telescope than with their small ones, and he invited Denning to come to Princeton to see for himself (unfortunately, he never did). He also tried to put the discussion into the proper perspective. "Life is short," he wrote, "and there are other interesting telescopic objects besides planetary markings."[26]

The discussion had begun in 1885 and had continued into 1886. By then, Henri Perrotin and Louis Thollon at Nice, using a 15-inch refractor there, had after many attempts finally succeeded in confirming the Schiaparellian canals, making them among the first professional astronomers to do so. Moreover, they had seen the geminations as well. The canals were gaining ground. Meanwhile, the debate over small versus large telescopes was to lapse for a few years, only to erupt again in the 1890s, with the canals once again at the center of the discussion. As for discoveries of planetary surface details of the sort that Denning was demanding from the large telescopes, the first of these was less than two years away, and it would be made with the newly unveiled 36-inch refractor at California's Lick Observatory. For the moment, the large telescopes had apparently been routed—though only apparently. In fact, Young's defense of them had been skillful enough. He was simply ahead of his time. When some of the issues raised by

Denning were to come up again a decade later, the terms of
the argument would be rather different; it would be the users
of small telescopes who would then be put on the defensive. In
fact, Young's role in the Martian canal debate, like that of
Green before him, has been unjustly forgotten. Yet it is a
matter of record that he was the first to realize what other
users of large telescopes were to rediscover in later years, and
knowing the sequel makes the following words seem prophetic
indeed:

> The discordance between the different maps of Mars indi-
> cates that the best and most keen-eyed observers uncon-
> sciously supplement what they really see, with details
> which they only *think* they see; so that in the finished
> drawing fact and fancy are inextricably mingled. The later
> observer with larger telescope and higher power naturally
> fails to recognize many features, and some, he has to
> repudiate.[27]

At the moment, however, which markings would be repudi-
ated and which would be confirmed if and when the large telescopes finally caught up with the small ones was far from
clear.

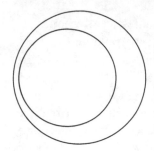

CHAPTER 10

# Confirmations?

*Illusions form the unstable basis of our
ideas, of our sensations, of our
sentiments, of our beliefs.*
　　　　　　　　—Camille Flammarion

AT FIRST Schiaparelli's drawings and
maps remained largely unconfirmed. Though a number of observers, especially in Britain, early reported *something* corresponding to the Schiaparellian canals, they had not agreed
with the way the Italian observer had interpreted these details.
It had been obvious to a number of them that they had been,
to put it bluntly, caricatured. As Webb had noted, Schiaparelli
had "plotted a sharply-outlined chart, which . . . no one would
at first imagine to be intended as a representation of Mars,"
having made use of an "unpleasantly conventional" style.[1]

Had Schiaparelli been less eminent in the field, the maps
might well have been dismissed without more ado. But his
reputation made it impossible to dismiss them out of hand,
and the tendency was perhaps rather to take the view of Otto
von Struve, who had supervised Schiaparelli's training at
Pulkova. Commenting on his own inability to confirm Schiaparelli's discoveries, Struve noted: "I am sorry to say that I
have never been able to see the canals, but knowing M.
Schiaparelli's excellence as an observer, I cannot doubt that
they are there."[2] Thus the fact that the canals had been seen by
such a skilled observer as Schiaparelli was for Struve a sufficient warrant of their reality, and it became a challenge to
other observers to find them "now that they knew they were
there." To fail to see them amounted—for observers with less
self-esteem, perhaps, than Struve—to an admission of observational obtuseness. Observers therefore proceeded to see, or
to think that they saw, the canals, the whole process bearing a
distinct resemblance to that described in the story of the emperor's new clothes. By the 1890s the canals were generally

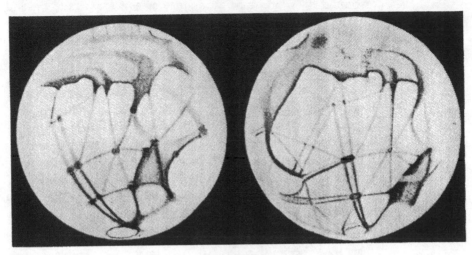

*Figure 10.1*
*Mars as drawn by Schiaparelli on*
*May 27 and June 2, 1888, showing*
*the highly geometric forms given to*
*the Martian surface markings in*
*Schiaparelli's later drawings. The*
*instrument used was the 18-inch*
*Merz-Repsold refractor of the*
*Brera Observatory. Following its*
*erection in 1886, this was*
*Schiaparelli's chief instrument*
*for planetary work, largely*
*replacing the celebrated 8.6-inch*
*Merz refractor.*

accepted by professional and amateur astronomers alike, and hardly a map of the planet was produced that did not show them in considerable profusion.

The analogy to the story of the emperor's new clothes is not quite precise. The Martian "deserts" were not altogether bare; there was *something* in the regions where Schiaparelli drew his canals. It proved only too easy for observers to convince themselves that this something had more or less the geometrical form Schiaparelli had given to it. As Rev. T. E. R. Phillips, a noted British amateur, wrote on the basis of observations with a 9½-inch reflector in 1898, "Faint markings have been glimpsed now and then, and it would be easy, by the 'scientific use of the imagination,' to conjure them into lines and streaks harmonising with Schiaparelli's charts."[3] Whatever else they may have been, the succession of maps of the planet—all showing canals, and many of these in the precise positions depicted by Schiaparelli—were remarkable testimonials to the powerful influence of suggestion.

Though he was referred to as the "prince of Mars observers" and not infrequently acclaimed the greatest astronomer of his century, Schiaparelli once wrote (more modestly and certainly more accurately) to Hector MacPherson, who was writing an article about him, that "in my estimation I found in the article a few exaggerations in my favour. The epithet of *great astronomer* is appropriate for such men as Hipparchus, Ptolemy, Copernicus, Kepler, Newton, Laplace, Bessel. My small labours do not give me the right to enter the ranks of such men."[4] On the other hand, in the Latin verses he wrote in dedicating the third volume in the series of his Martian memoirs to Professor Tito Vignoli in 1886, he was more grandiloquent and left no doubt that he considered his canals a discovery of the first order of importance. His poem reads in part:

> The truth about the stars Urania brings,
> And from our own Milan the glory springs,
> New features showing on the face of Mars
> As it wheels endlessly among the stars:
> Behold there worlds unknown to earthborn eyes,
> And what new wonders issue from the skies!

Parallel furrows in the ruddy sphere,
Drawn by some unknown Daedal's hand, appear.
Beside Thaumasia shines a land of gold,
Whose ample shores jewel-bearing seas enfold;
Ausonia's wide expanses glimmer there,
And seas Tyrrhenian gleam in azure air;
The Syrtes, feared by sailors in old time,
Lie torpid 'neath the Titan's weltring clime,
And flowery shores the fertile Ganges laves,
Gladdening each bank with its genial waves.
'Mid polar night does Thule hug the snow,
And Acheron black, and famous Simois, flow.
There streams of milk, and fonts of youth, are found;
And honey sweetens Hybla's sacred ground.[5]

Those who had seen Schiaparelli's maps could hardly approach the planet in an unprejudiced manner, and most hardly tried. François Terby, a Belgian astronomer who wrote a large book on Mars, used an 8-inch reflector and, as many did, took Schiaparelli's map with him to facilitate identification of the canals:

We took inspiration from the principle announced by some great observers: "Often," they say, "one can see well what one especially seeks." . . . We have sought the canals of Mars in the regions where we knew that M. Schiaparelli had proved them to exist . . . and, map in hand, we have patiently and obstinately pursued these very difficult details. It is to this method . . . we owe our partial success. The observer armed with a simple 8-inch in the unfavorable circumstances which we have experienced, would be discouraged very quickly if he were not sustained in advance with an unbreakable faith in the truth of the Milan results; this faith alone, in effect, could inspire him with the perseverance and, I may say, the obstinacy necessary. Furthermore, prodigious difficulties are a great help to incredulity, and it is permissible to believe that observers provided with adequate means . . . have often abandoned the observation through lack of preliminary confidence or of anterior knowledge of the details sought for.[6]

Commenting on the fashion for canals on maps of the 1890s, even Percival Lowell had to admit that

> we easily see what we expect to see, but with great diffi-
> culty what we do not. This may be due to individual idio-
> syncrasy, or it may be due to a prevailing idea of the time,
> affecting people generally, in which we unwittingly share.
> Fashion is as potent here as elsewhere. The very same
> cause will show us at one time what we remain callously
> blind to at another. A few years ago it was the fashion not
> to see the canals of Mars, and nobody except Schiaparelli
> did. Now the fashion has begun to set the other way, and
> we are beginning to have presented suspiciously accurate
> fac-similes of Schiaparelli's observations.[7]

One would be hard pressed to find a more cogent account of the sociology of the Martian canal episode; yet even possessed of such insight, Lowell was himself unable to rise above the fashion of the times, being one of those who was most unwittingly affected by the prevailing idea of the canals.

To give Terby and others their due, what we have called preperception can greatly assist detection, as we have already discussed in some detail. Nothing is truer than the idea behind Sir William Herschel's maxim: "When once an object is seen with a superior telescope, an inferior one will suffice to see it afterwards."[8] Terby seemed to be paraphrasing this remark when he said, "Often one can see well what one especially seeks." Yet undoubtedly this factor of "preliminary confidence"—or "anterior knowledge," as Terby put it—while a powerful weapon, can be a double-edged one. Here some comments by Sir Ernst Gombrich are pertinent, though in the case of planetary perceptions the "whiffs" are not of sound but of visual information:

> I was employed for six years by the British Broadcasting
> Corporation in their "Monitoring Service," or listening
> post, where we kept constant watch on radio trans-
> missions from friend and foe. It was in this context that
> the importance of guided projection in our understanding
> of symbolic material was brought home to me. Some of
> the transmissions which interested us most were often
> barely audible, and it became quite an art, or even a sport,

to interpret the few whiffs of speech sound that were all we really had on the wax cylinders on which these broadcasts had been recorded. It was then we learned to what extent our knowledge and expectations influence our hearing. You had to know what might be said in order to hear what was said. More exactly, you tried from your knowledge of possibilities certain word combinations and tried projecting them into noises heard. The problem was a twofold one—to think of possibilities and to retain one's critical faculty. Anyone whose imagination ran away with him, who could hear any words—as Leonardo could in the sound of bells—could not play that game. You had to keep your projection flexible, to remain willing to try out fresh alternatives, and to admit the possibility of defeat. For this was the most striking experience of all: once your expectation was firmly set and your conviction settled, you ceased to be aware of your own activity, the noises appeared to fall into place and be transformed into the expected words. So strong was this effect of suggestion that we made it a practice never to tell a colleague our own interpretation if we wanted him to test it. Expectation created illusion.[9]

There is no other explanation than Gombrich's that "expectation created illusion" for the fact that Herschel, who at first thought Uranus a comet, found its diameter increasing steadily in the period after its discovery, as would have been expected of a comet approaching Earth (in fact, the actual diameter was decreasing slightly over the time involved).[10] There is no other explanation for the supposed resolution of the Dumbbell and Orion nebulas into stars by Lord Rosse and G. P. Bond before the spectroscope of Sir William Huggins showed conclusively that they were, in fact, not constituted of stars (as a number of other nebulas had been shown to be) but of diffuse gas.[11] In the same way, expectation to a large extent created illusion for many of the post-Schiaparellian observers of the canals of Mars.

The details of the constantly fluctuating planetary disk became sharp for only brief interludes and then blurred out. They had to be actively sought out. In order to record anything at all, the observer had to be prepared and selective because this very selectiveness was what allowed for the immediacy of perception that enabled one to follow the moment-to-moment comings and goings of fine details. The photographic plate was, as we have seen, too slow to do so; it could catch only the coarser features of a planet's disk, not those that lay at the threshold, which is, after all, the region of discovery. Moreover, while the camera's recorded image could be studied later, the visual observer had to bring his hypothesis directly to the telescope and test it there.

Thus the visual observer's record bore the inevitable stamp of subjectivity, being not an impartial recording of the surface of the planet, an objective datum, but rather an impression. The data were not recorded impartially and interpreted at one's convenience later; the interpretation was inextricably and unavoidably melded together with the recording of the data itself. Of course, objectivity is the goal to which science aspires, but in the field of visual planetary observation, it seemed defeated at the outset. The real question seemed to be whether the question was not, like those considered in introspective psychology, so entirely subjective as to be hardly within the proper domain of science at all. What one man affirmed, another could—and did—as easily deny, and each seemed to hold absolute sovereignty over his own visions. Lowell said of Schiaparelli: "He perceived what he saw, which is where most persons fail."[12] But what he saw was not necessarily objectively real. As usual when a matter is capable of neither final proof nor final disproof, it became, as often as not, a matter of faith. Terby, as we have seen, spoke of having "unbreakable faith in the truth of the Milan results," and Lowell was to say of those who doubted the canals' existence: "The spirit that denies has always been abroad; only in early days he was reputed to be the devil."[13]

By the 1890s the canals, having begun with Schiaparelli's private visions, had become, it is perhaps fair to say, a "collective illusion." Much was made of the apparent "verification" made by the French astronomers Henri Perrotin and Louis Thollon at Nice Observatory using the 15-inch refractor there in 1886. These observers gave the following descripton of the planet as it appeared to them with magnifying powers of 450 to 560×: "In the equatorial regions [is] a network of lines running in various directions, and projecting themselves upon the brilliant ground of the disk, as greyish lines of color more or less intense."[14] Percival Lowell later pronounced these observers the first to see the canals after Schiaparelli himself (though this is certainly incorrect, as we have seen that the British observers Burton and Maunder had reported them as early as 1879 and 1881). But in any case, there can be no doubt that Perrotin and Thollon had been actively pursuing the canals. "In spite of the great size of the glass," Lowell wrote,

> a first attempt resulted in nothing but failure. So, later, did a second, and Perrotin was on the point of abandoning the search for good, when . . . he suddenly detected one of the canals, the Phison. His assistant, M. Thollon, saw it immediately afterward. After this they managed to make out several others, some single, some double, substantially as Schiaparelli had drawn them; the slight discrepancies between their observations and his being in point of fact the best of confirmations.[15]

Perrotin continued to draw canals in 1888, now having the advantage of the 30-inch refractor Lowell had mistakenly assumed he had used at the previous opposition. This telescope was constructed with the unusually long focal ratio of f/24 specifically with planetary work in mind.[16] Perrotin reported witnessing the inundation of the continent that Schiaparelli called Libya by the neighboring sea ("if sea it is," Perrotin added in a prudent afterthought). Unfortunately, observers equipped with a still more powerful telescope, the recently erected 36-inch refractor on Mt. Hamilton, in California, had also been watching Mars and had found no change in the region.

THE ERECTION of the great Lick refractor had been made possible through the bequest of real-estate millionaire James Lick, who had won his fortune through shrewd investments during the California Gold Rush.[17] At first, Lick entertained a plan to use some of his fortune to build the world's largest pyramid as a monument to himself (it was to be located on the corner of Fourth and Market streets in San Francisco). By

1873, the year of the completion of the great refractor of the U.S. Naval Observatory in Washington (and possibly as a result of the publicity given that event), his imagination was stirred by the idea of providing funds to construct a telescope "superior to and more powerful than any yet made."[18] Originally this, too, was to be situated in downtown San Francisco; fortunately, he was later persuaded that the telescope would have more value astronomically if it were set up in a more favorable site. A site in the High Sierras, near Lake Tahoe, was considered for a time, but Mt. Hamilton, near San Jose, was eventually chosen.

Not surprisingly, Alvan Clark was placed in charge of building the telescope, but there was a delay before he could begin work, as it proved difficult to obtain large glass disks of sufficient quality for the lens. It was not until 1885 that suitable disks were delivered. By the fall of the next year, however, the lens had been finished in the Clarks' Cambridge workshop and shipped to San Jose. The building of the mount for the telescope took another year. It was not until New Year's Eve in 1887 that the lens was finally mounted in its tube by Alvan Clark's son Alvan Graham Clark. By that time Alvan Clark himself was dead, as was James Lick, whose body was interred in the pier of the great telescope itself.

In what was practically the maiden view with the telescope, James E. Keeler, on January 7, 1888, detected a fine division in the outer ring of Saturn (Fig. 10.2). This was not the so-called Encke division that had often been reported by previous workers (among others, by the onetime director of the Berlin Observatory Johann F. Encke). In the location of the Encke division Keeler saw a broad shaded area, but in addition there was a much more sharply defined, but exceedingly narrow, division farther out on the ring. Keeler afterward described this as "so minute that it may fairly be classed among the most difficult and delicate of planetary details, requiring the most powerful instruments and exceptional conditions for its observation."[19] In fact, it proved possible to use magnifications as great as 1500× with the new instrument under the best conditions, and a planet might appear with "a sharpness more characteristic of the lines of a steel engraving than of the usual telescopic image," as Keeler wrote on a subsequent occasion.[20] Keeler's detection of the new fine division in the outer ring of

*Figure 10.2*
*Saturn as drawn by James E.*
*Keeler with the 36-inch refractor*
*of the Lick Observatory, January 7,*
*1888, one of the first observations*
*with the great telescope. It shows*
*the narrow gap near the outer*
*edge of Ring A, which Keeler*
*was the first to describe. Though*
*difficult to reproduce, the original*
*drawing, located at the Yerkes*
*Observatory, also gives hints of the*
*"spokes" on the surface of Ring B,*
*which were conspicuous in the*
Voyager *spacecraft photographs*
*of 1980 – 81.*

Saturn went completely unheralded. This was unfortunate, since Denning's challenge of 1885 to users of large instruments—that they prove the merit of such instruments by discovering some new marking on a planetary surface—had finally been answered.

Though it proved its worth on Saturn, the Lick refractor achieved only mediocre results when turned on Mars. Not until July 1888 were the first observations of that planet made, the telescope being shut down for a number of months after its first tests until the observatory officially opened. Mars was then some three months past what had been an only average opposition and showed an apparent diameter of only 9″ of arc. Moreover, it was far to the south, never rising more than 30° above the horizon (compared to Saturn, which had been near the zenith); it was therefore impossible to use such high magnifying powers as had been used on the ringed planet.

In any case, it was soon obvious that the Lick observations would be just as distractingly subject as any others to the "personal equation" of the observer, which made the question of whether changes on the surface of Mars had taken place seem no more likely to be resolved by visual observations than the analogous question concerning the Moon in the days of Schröter and Gruithuisen. The representations of Holden and Keeler, for example, differed not only from those of Schiaparelli and Perrotin but also markedly from one another, prompting Camille Flammarion's riposte:

> It is impossible to look at these views of Mars . . . with any attention at all and not be struck at once by the differences. For example, the drawings by Holden and Keeler, being made at the same time with the same instrument, . . . can only be supposed to differ in the observers; for no one would have the temerity to imagine that the planet itself could have undergone such a metamorphosis in a quarter of an hour! . . . Such marked discrepancies must make one very wary indeed of reaching any conclusions about variations on the planet's surface. . . . Can one actually suppose that it is even the same face of the planet that is depicted here? . . . What difference of aspects![21]

That the planet should appear in different guise to different observers was only, perhaps, to be expected, but even the same observer could be baffled by the "difference of aspects." Schia- parelli, writing in this same year of 1888, on one occasion noted that "the canals had all the distinction of an engraving on steel with the magical beauty of a coloured painting";[22] but on another occasion he exclaimed:

> What strange confusion! What can all this mean? Evi-
> dently the planet has some fixed geographical details, sim-
> ilar to those of the Earth. . . . Comes a certain moment,
> all this disappears to be replaced by grotesque polygona-
> tions and geminations which, evidently, attach themselves
> to represent apparently the previous state, but it is a gross
> mask, and I say almost ridiculous.[23]

This was the closest that the "father" of the canals would ever come to disowning his strange progeny.

BY THE OPPOSITION of 1890, Mars was considerably nearer to the Earth than it had been in 1888, but it was still farther south, and consequently the planet could be observed even less satisfactorily from the northern hemisphere. Nevertheless, Holden at Lick sanguinely reported that "the positions of most of Professor Schiaparelli's canals have been verified by some one of us."[24]

The opposition of 1892 was eagerly awaited, since Mars would then approach within the 35-million-mile mark for the first time since 1877. An interesting series of observations was reported in 1892 by Young at Princeton with its 23-inch Clark refractor:

> Of course I tried very earnestly to make out the "ca-
> nals," which figure so conspicuously in Schiaparelli's maps,
> but mostly without success. There were, indeed, various
> faint markings some of which with a lower power seemed
> to correspond fairly in position and general direction with
> "canals" shown upon the map; but under higher magnify-
> ing powers the resemblance disappeared; that is, instead
> of being narrow lines well defined and nearly straight,
> they became mere shadings, irregular, indefinite and
> vague in outline, and often discontinuous.[25]

As already noted, Young preferred Green's representation of the Martian surface to Schiaparelli's; yet, while skeptical of what the Italian astronomer had seen, he did not deny outright the reality of the canals, adding that while in 1892 seeing was "unusually fine for Princeton . . . [it] was probably not equal to that which prevails in the Italian atmosphere, nor can I pretend to any remarkable keenness of vision."[26]

Holden, at Lick, gave a general report of observations made in 1892 at Mt. Hamilton. Once again the planet's far southerly position made it difficult to make use of the highest magnifications; Holden noted that "most of the drawings with the 36-inch have been made with a magnifying power of 350 diameters; a few with the power of 260. . . . How unfavorable the circumstances have been can be estimated when it is remembered that powers of 1000 and even more have been employed on Jupiter and Saturn with good result."[27]

Other reports of interest included Pickering's detection at Arequipa, Peru, of a series of "lakes" occurring at the intersections of the canals, of which more will be said later, and Perrotin's observations of "clouds" projecting beyond the planet's limb (similar phenomena had been seen at Lick in 1890 and by Terby in Belgium in 1888). There were a few in the press who were quick to declare that these clouds were actually attempts of intelligent Martians to communicate with Earth.[28] Far-fetched as the idea was regarded even at the time, it does serve as an indicator of the popular mood about Mars even before Percival Lowell arrived to stir it up further.

For the most part, however, the 1892 opposition was a disappointment; it did not lead to developments of comparable significance to those of the previous close approach of 1877, when the canals and the satellites had been discovered. Indeed, as Clerke wrote in her *Popular History of Astronomy*, "The low altitude of the planet practically neutralised [the advantage of distance] for northern observers, and public expectation, which had been raised to the highest pitch by the announcements of sensation-mongers, was somewhat disappointed at the 'meagreness' of the news authentically received from Mars."[29] Yet it was perhaps significant that observers using the largest telescope in the world had apparently confirmed Schiaparelli's findings, and Clerke concluded that "the 'canals' of Mars are an actually existent and permanent phenomenon."[30]

It may be, though, that the most significant development
during the 1892 opposition was actually the entry into the
field of the skilled observer E. E. Barnard, who was given the
opportunity to use the great Lick refractor for the first time
that year. If one can believe what a later colleague of his said
concerning what he saw that year, it was nothing short of
remarkable—and it was of a significance far beyond all the
other discoveries of that opposition combined.

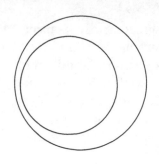

CHAPTER 11

# The Sensitive Eye

*Who then shall teach thee, unless it be thine own eye?*

—*Euripides,* Helen

THOUGH HE HAD been on the Lick staff for four years, not until August 1892, the month of Mars's closest approach to the Earth, did Edward Emerson Barnard finally receive permission to use the 36-inch refractor regularly for his observations (previously he had been allowed use of only the observatory's 12-inch refractor). Having failed in several attempts to persuade director Holden of his worthiness for the larger instrument, Barnard at last took his case directly to the regents of the University of California. His request was granted, and he took charge of the great telescope every Friday night. Within a month, he had achieved international fame by discovering Amalthea, the difficult inner satellite of Jupiter (and the last satellite in the solar system, by the way, to be discovered visually).[1]

Barnard, a native of Nashville, Tennessee, was in every respect a self-made man, rather in the mold of Alvan Clark.[2] His father died before he was born, and Barnard was educated at home by his mother. At the age of nine he went to work in a photographer's studio. Mundane as many of his responsibilities were on the job—among other things, he had to keep a large solar camera used to make photographic enlargements pointed toward the Sun—he was later to make good use of what he learned in the wide-field photography of the Milky Way, which he pioneered.

He became seriously interested in astronomy at a comparatively late date. At the age of nineteen he happened across a copy of Dr. Thomas Dick's *Practical Astronomer;* he read the book as a revelation. Not long afterward he was making his

*Figure 11.1*
*Edward Emerson Barnard*
*(1857 – 1923).*

first telescopic observations. Initially he made do with a makeshift telescope fashioned from the lens of a broken spyglass he is said to have found in the street; but this modest instrument did not satisfy him for long, and soon he and his mother were reaching into their savings so that he could acquire a better telescope, one with a 5-inch aperture. At about the same time, he had the opportunity to meet the famous astronomer Simon Newcomb, who when asked by the young man how he might contribute to astronomy told him to look for comets and get a thorough grounding in mathematics.

On one night in May 1881, Barnard discovered a comet, observed it again the following night, but subsequently lost sight of it. He was therefore denied credit for the find. This was more than a loss of prestige. A prize of two hundred dollars (a fairly substantial sum in those days) had been offered by a New York patent-medicine vendor, H. H. Warner, for every new comet discovered by an American, and Barnard was then in rather desperate financial straits. Fortunately, Barnard was able soon afterward to find several other comets, winning enough money thereby to build a house in Nashville for the wife he had married in January 1881 and for his mother, who was now an invalid; fittingly enough, it became known as Comet House.

Barnard became well known for his comet discoveries and other observations, especially of Jupiter, and on this basis he was recommended for a fellowship at Vanderbilt University, where he could at last follow Newcomb's advice that he study mathematics. In addition, he took charge of the observatory there. Though he did not finish his bachelor's degree until he was thirty, his work at Vanderbilt was so impressive that he received an invitation from Holden, with whom he had been corresponding, to join the staff being organized at Lick. He arrived at Mt. Hamilton on May 1, 1888, and remained there for seven years, finally departing in 1895 to take a position at the new Yerkes Observatory of the University of Chicago, whose 40-inch refractor was to surpass even the great Lick instrument.

As an observer, Barnard pushed himself relentlessly. At Yerkes, where he had two regular observing nights per week with the 40-inch, he became known for his pacing and ner-

vous coughing if one of his nights approached with the prospect of overcast skies. E. B. Frost, director of Yerkes when
Barnard was there, said that it was one of his most serious
duties to keep Barnard from working himself too hard, while
Barnard himself, who suffered from diabetes in later years,
recalled that one of his greatest trials was to have to refrain for
an entire year from use of the 40-inch on orders of his doctor.
One can only imagine how upsetting it must have been for
him at Lick before he received regular use of the 36-inch
refractor, for he cannot have failed to notice, while working on
the humble 12-inch, that Holden, who reserved the great telescope for himself *two* nights a week, often shut down the
dome with the sky still clear and went to bed after only a few
hours of work.

WITH THIS INTRODUCTION, we turn to Barnard's observations of Mars, which have occasioned a fair amount of interest in recent years because of a report by a colleague of his at
Yerkes, John E. Mellish, that Barnard saw the Martian craters
during his tenure at the Lick observatory. These were first
officially discovered in the Mariner 4 spacecraft photographs
of July 1965. Mellish himself claimed to have seen them in
November 1915 with the Yerkes 40-inch. As he wrote to Walter
Leight in 1935, he had been observing the planet in an exceptionally steady atmosphere just after sunrise and noted, with
magnifications of 750× and 1100×, "something wonderful
about Mars. It is not flat, but has many craters and cracks. I
saw a lot of the craters and mountains one morning with the
40-inch and could hardly believe my eyes."[3] When Mellish told
Barnard what he had seen, Barnard fetched some of his own
drawings from an old trunk, drawings that, according to
Mellish, had been made with the Lick 36-inch in 1892–93.
Mellish called these "the most wonderful drawings that were
ever made of Mars." They showed, he recalled,

> the mountain ranges and peaks and craters and other
> things, both dark and light, that no one knows what they
> were. I was thunderstruck and asked him why he had
> never published them. He said, no one would believe him
> and [others] would only make fun of it. Lowell's oases are
> crater pits with water in them, and there are hundreds of

brilliant mountains shining in the sunlight. Barnard took
whole nights to draw Mars and would study an interesting section from early in the evening when it was coming
on the disk until morning, when it was leaving. He made
the drawings four or five inches [in] diameter, and it is a
shame that those were not published.[4]

It is indeed a shame that Barnard's drawings were not published, and until very recently they were generally believed to
have been lost. Their actual fate is discussed below. Mellish's
drawings of 1915–16 were never published and were lost in a
fire—ironically, the year before the Mariner 4 flyby.[5]

Barnard himself had nothing to say about either craters or
mountains in his own preliminary remarks on his observations
of 1892. Instead, he mentioned only his failure to see any
geminations of the canals. He did see a few of the canals
themselves, as his published drawings show. But then, observing conditions that year were very poor, as Barnard told
Schiaparelli during a visit to Milan in the summer of 1893.
Because of its lowness in the sky, Mars did not allow the use of
high power; Barnard's best results were obtained with only
$200\times$ on the 12-inch and $350\times$ on the 36-inch.[6]

Conditions were vastly better in 1894, with Mars higher in
the sky as seen from Mt. Hamilton. If Barnard ever saw the
Martian craters, it must have been that year (rather than
1892–93, as Mellish recalled), for the views of Mars with the
36-inch were indeed wonderful. As he wrote to Newcomb,

I have been watching and drawing the surface of Mars. It
is wonderfully full of detail. There is certainly no question
about there being mountains and large greatly elevated
plateaus. To save my soul I can't believe in the canals as
Schiaparelli draws them. I see details where some of his
canals are, but they are not straight lines at all. When best
seen these details are very irregular and broken up—that
is, some of the regions of his canals; I verily believe—for
all the verifications—that the canals as depicted by Schiaparelli are a fallacy and that they will be so proved before
many oppositions are past.[7]

Though he confided the results of his 1894 Martian observations privately to Newcomb, Barnard's first published report of

what he had seen did not appear until two years later, when he made reference to his work on Mars with the great refractor almost incidentally in a discussion of his 1894–95 Saturn observations. He had felt obliged to address "the apparently abnormal lack of details shown in my observations of Saturn with the 36-inch" compared to those of some observers using much smaller instruments. This lack of details, he said, had led to the impression that, for various reasons, "great telescopes are inferior to smaller ones for showing the delicate markings on the surface of a planet."[8]

The subject of small versus large telescopes had, of course, been ably raised by W. F. Denning in the mid-1880s, and indeed in 1895 Denning himself had published yet another article in which—though still advocating smaller telescopes—his enthusiasm was somewhat tempered:

> The phenomenal results recently claimed for certain small telescopes are almost of a character to shake the faith of those disposed to acknowledge their great utility on several classes of objects, for our confidence cannot go beyond reasonable limits. In individual cases a good though small instrument, an acute well-trained eye, acting in combination with the best atmosphere and conditions, will yield surprising results; but some of those lately published border upon romance, and henceforth it would seem that if all the data described with such means are to be absolutely accepted, then large telescopes are grossly incapable on certain important objects, and may as well be packed away in the lumber rooms of our observatories.[9]

Among the observations that clearly strained Denning's credulity were those of spots on Saturn made by a British amateur, Arthur Stanley Williams. With only a 6½-inch reflector, Williams had seen some of the spots even when the planet was nearing conjunction with the Sun and low down, near the horizon—both unfavorable circumstances for such observations. A few observers provided corroborations, but this hardly eased Denning's doubts, for he noted: "When we consider that many hundreds of amateurs have been employing their telescopes upon Saturn without seeing the spots, the affirmative evidence of a few isolated persons can hardly be regarded as conclusive. It is a fact that, if any new feature on a planet . . . were confidently announced, a few of the many observers who looked for it would certainly assert they could see it though not really existing."[10]

Denning, with his 10-inch reflector at Bristol, had been one of the hundreds of amateurs who had been employed on Saturn without success, and he offered the following advice to observers:

> In a case of this kind the observer has to be severe with himself. There is a distinct line of demarcation between what is absolutely seen and what is possibly seen or suspected. An object may only be glimpsed, and yet it is certainly seen, for its impression reaches the eye now and then in a form not to be mistaken. But with some objects the experience is different. We fancy that they are there, but cannot fix them with certainty; apparently they flit about like an *ignis fatuus,* and are intractable to our utmost efforts. Obviously in such a case the observer has but one alternative, . . . to regard the objects as imaginary.[11]

C. A. Young had put the matter even more bluntly. Small images, he had said, "are very encouraging to the imagination."[12]

Williams's testimony was not, however, to be dismissed so lightly. A solicitor by profession, he was a leading observer in the recently founded British Astronomical Association (BAA). His career as an astronomer had begun in 1877, when he was only sixteen. When everyone else's eyes were turned toward Mars, Williams began a special study of Jupiter that was to continue for many years. He devised the method of timing the transits of the various features which has remained, essentially, that of the Jupiter section of the BAA ever since, and he used it to establish the existence of nine currents in the Jovian atmosphere by 1896.[13]

It was natural that Williams should eventually turn his attention to Saturn, in many ways a kindred planet to Jupiter, and in 1893 he announced that he had succeeded in detecting a number of light and dark spots on the planet.[14] Barnard, after observing Saturn carefully with the 36-inch refractor in

1894, found the spots "totally beyond the reach of the 36-inch . . . as well as of the 12-inch under either good or bad conditions of seeing."[15] On the drawing he published with his report he commented caustically: "I have only drawn what I have seen with certainty. It is true that the picture appears abnormally devoid of details when compared with drawings made with some of the smaller telescopes. I am satisfied, however, to let it remain so."[16]

Denning had urged that the debate between large and small telescopes be settled by careful trials of the instruments side by side, and added: "If observers having the appliances at command will institute some further comparisons of the kind suggested, the problem might be virtually solved in a short time."[17] Barnard was interested enough to perform a few experiments with diaphragms, similar to those performed by Young at Princeton a decade earlier. He viewed both Venus and Saturn on several occasions, with the full aperture and with the aperture diaphragmed to 12 inches. Barnard concluded: "I am convinced that everything that can be seen with the telescope diaphragmed down can be better seen with the full aperture when the air is steady. If the object is very bright, or the air unsteady, I think a reduction of aperture would be an improvement."[18]

Though the debate over small versus large telescopes did not end with Barnard's experiments, certainly Barnard's contribution was such that no one who subsequently took up the cudgel, on one side or the other, could safely ignore it. More impressive than his experiments with diaphragms, however, were his detailed descriptions of his own planetary observations with the great Lick refractor. Of Jupiter he wrote: "The view of this splendid planet, with this noble instrument, under first-class conditions, is magnificent, and the amount and intricacy of detail is utterly beyond the ability of an observer to depict."[19] At last he turned to his 1894 observations of Mars. Beginning with his skepticism toward the drawings of observers using small telescopes—"If we are to take the testimony of the drawings themselves," he wrote, "the smaller the telescope the more peculiar and abundant are the Martian details"—he went on to describe the quite different class of features brought out by the 36-inch refractor in 1894:

On several occasions during that summer, principally when the planet was on the meridian shortly after sunrise—at which time the conditions . . . are often exceptionally fine at Mount Hamilton—its surface with the great telescope has shown a wonderful clearness and amount of detail. This detail, however, was so intricate, small, and abundant, that it baffled all attempts to properly delineate it. Though much detail was shown on the bright "continental" regions, the greater amount was visible on the so-called "seas." Under the best conditions these dark regions, which are always shown with smaller telescopes as of nearly uniform shade, broke up into a vast amount of very fine details. I hardly know how to describe the appearance of these "seas" under these conditions. To those, however, who have looked down upon a mountainous country from a considerable elevation, perhaps some conception of the appearance presented by these dark regions may be had. From what I know of the appearance of the country about Mount Hamilton as seen from the observatory, I can imagine that, as viewed from a very great elevation, this region, broken by canyon and slope and ridge, would look just like the surface of these Martian "seas." . . . At these times there was no suggestion that the view was one of far-away seas and oceans, but exactly the reverse. Especially was I struck with this appearance in the great "ocean" region of the Hour-glass Sea [Syrtis Major], and especially in the equatorial portion of this region. These views were extremely suggestive and impressive. I have not seen these small and delicate details described elsewhere, and I feel confident they would scarcely be shown in a much smaller telescope. The details shown on the "continental" regions were usually irregular features, principally delicate differences of shade. No straight hard sharp lines were seen on these surfaces, such as have been shown in the average drawings of recent years.[20]

Such, then, were Barnard's observations of 1894. He seems explicit enough about what he did see, yet there has been, ever since Mellish published his comments, the lurking suspicion that Barnard never made his most interesting results public. One could only hope that "the most wonderful drawings of Mars ever made," as Mellish called them, would turn up one day.

The logical place to look for them would be at Yerkes, where Mellish claimed to have seen them. I learned from Dale P. Cruikshank, who had done some research in the archives there, that he found a number of notebooks Barnard had made while at Lick. These were actually copies of Barnard's notebooks; Barnard had employed a young man to make them in 1895 when he was preparing to leave Mt. Hamilton, because Holden had at first refused to grant him permission to take the originals with him. Later Holden relented, so that the copies never actually had to be used. They remained at Yerkes (the originals, once Barnard had finished with them, were sent back to Lick).

This was encouraging, and it seemed worthwhile to look through the Yerkes archives again for Barnard planetary drawings. I am glad to report that the drawings were duly found, including some of Mars that had been gathered into bundles and had long been neglected. Of these, a few were from 1892, but the best by far were from 1894. They were drawn on a scale of five inches to the diameter of the planet, just as Mellish remembered, and a few of them had actually been published—some of the 1894 drawings on an unfortunately small scale in the 1895 edition of Chambers's *Story of the Solar System* and one of these on a larger scale in the 1903 *Astrophysical Journal*.

Perhaps the most spectacular of all the drawings is the one of September 2, 1894 (Fig. 11.2), which shows the four great volcanic calderas Olympus Mons, Arsia Mons, Pavonis Mons, and Ascraeus Mons with breathtaking detail. They are not shown as mere ink-black spots, as observers were wont to represent the so-called oases; instead, the forms in Barnard's drawings suggest true elevations of the surface, the dark kernels at their centers giving at least a hint of the calderas that are now known to cap their summits.

It is obvious from Mellish's descriptions that his memory of Barnard's drawings was imperfect in various respects, yet there seems no reason to doubt that these are the drawings he had

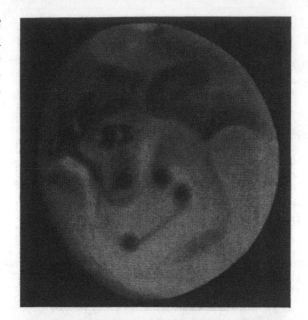

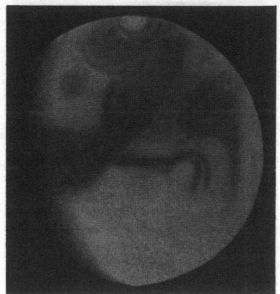

*Figure 11.2*
*Mars in 1894, as sketched by E. E. Barnard with the 36-inch refractor of the Lick Observatory. The upper sketch, made on September 2, 1894, shows the great vol-canoes of the Tharsis region. Note the irregularities of the termina-tor in the lower sketch, suggesting unevenness of terrain.*

in mind. (He embellishes a bit, no doubt, though his comment that there are "crater pits with water in them" or "hundreds of mountains shining in the sunlight" refer not to Barnard's observations but to Mellish's own, as he clarified before his death in 1970.)

Together with E. M. Antoniadi's observations with the 32³/₄-inch refractor at the Meudon Observatory in 1909, Barnard's views of Mars are undoubtedly among the most remarkable of the pre-spacecraft era. Had they become more widely known, with their eerie depiction of a Mars that looks like a dead and dreary wasteland of burnt-out volcanoes, one wonders whether Percival Lowell's theories would ever have been deemed credible.

There can no longer be any doubt that Barnard did record Martian craters—volcanic ones, at any rate—yet it would be misleading to suggest that he anticipated the spacecraft discoveries in any real sense. He never referred to the features that he sketched as craters, any more than Schiaparelli did—who, after all, first recorded Olympus Mons and one or two other features now known to be craters. It is possible that Mellish's observation of 1915 made Barnard reconsider the significance of what he had seen; Mellish's comments would seem to imply this, but we may never know for certain. Barnard's own observations from 1915–16 show nothing unusual, though it is only fair to point out that they were made with the 12-inch rather than the 40-inch refractor of the Yerkes Observatory.

Certainly after leaving Lick for Yerkes because of difficulties in dealing with Lick's director, E. S. Holden, in which he was not alone, Barnard never again obtained views of Mars like those of 1894. The atmospheric conditions in Wisconsin, where the 40-inch refractor was, as Percival Lowell put it, buried,[21] did not approach those on Mt. Hamilton, and in fact Barnard made relatively few visual observations of the planets there (though he did make some important contributions to planetary photography).

A vignette of Barnard at work at Yerkes at the 1909 opposition of Mars, when the planet was actually somewhat closer to the Earth than it had been in 1894, is of interest. James H. Worthington, a British amateur who visited a number of observatories that year, found Barnard measuring the positions of the Martian satellites with the 40-inch, and wrote: "I was amazed to find that it was only with considerable difficulty that I could see the satellites at all, and so difficult were they that I despaired of ever being able to set a wire upon them." Barnard, on the other hand, was obtaining a fairly concordant series of measures, "in spite of the troubled state of the atmosphere and the tremendous glare of the planet." As for details of the planet's surface, Worthington looked in vain for the set of delicate markings he had seen with Percival Lowell at Flagstaff a few weeks before, noting that the weather at Yerkes "was hopelessly unsteady for such work, and the aperture too great." Worthington, alas, had come on a poor night, and indeed one often encounters in Barnard's own notebooks comments such as, "The air is like running water in front of the planet," or "the planet is a mass of blurred light—there is no definite image."[22]

In any case, with Barnard unable to see at Yerkes what he had seen at Lick, the canalists, led by Percival Lowell, mounted a counterattack. Lowell was to prove a particularly persistent advocate of the use of diaphragms with large telescopes, and indeed insisted that "large telescopes are not always to be preferred for planetary work. On the contrary, small ones are not only sometimes but nine times out of ten more powerful in planetary visual research, strange as the fact may seem, than large ones." Interestingly, Barnard's own views about the usefulness of diaphragms mellowed after he left Mt. Hamilton. For instance, in a note concerning an observation of some spots he made out on Mercury with the 40-inch in August 1900—they reminded him of those seen on the Moon with the naked eye—he wrote: "The markings could be seen for a moment beautifully and I think would have been steadily with a diaphragm." (The diaphragm on the 40-inch had been taken off the objective at the time, he explained.) And later, as a member of a committee formed to make recommendations for planetary observers he endorsed the use of diaphragms on large instruments.[23]

Williams's work on Saturn was also being rehabilitated. E. M. Antoniadi—who was, ironically, later to become one of the leading exponents of the superiority of large apertures in planetary work—verified Williams's spots of Saturn with a 9½-

inch refractor in 1896−97 and went on record as agreeing with Lowell and others that Barnard was not particularly adept "as an observer of delicate planetary details."[24]

Thus, in Antoniadi's view at least, Williams's work with the small telescope was borne out over Barnard's negative testimony with the large one. Nor would it be at all fair to conclude that the users of small instruments made out *only* spurious details, as Barnard's comments might lead one to suppose. Even on Mars this was not the case. Though the average user of a small telescope was guilty of drawing hard, sharp lines on the planet of the sort to which Barnard objected, the recognition of true structure on the planet's surface was not quite the exclusive purview of the Lick 36-inch or the Yerkes 40-inch refractors. W. F. Denning proved that his faith in small telescopes was not entirely misplaced when he used his modest 10-inch reflector in 1886 for some remarkable observations of his own:

> The surface markings of this planet are so numerous and varied that they are far from being adequately represented on existing charts. In certain regions the disk is so variegated as to give a mottled appearance. Dark lines, and spots, and bright spaces are so thickly interspersed, and so difficult to observe with sufficient steadiness to estimate their positions and forms, that I found it impossible to make a thoroughly satisfactory drawing.[25]

Denning concluded that one had no choice but to content oneself with depicting the broader features only. As for the Schiaparellian canals, while admitting that there was a mass of complicated details present on the disk, Denning felt that these existed "scarcely in the form and character" under which they had been represented by the Italian observer.

Such observations show that, important as aperture might be, a talented observer could still make up for much of the disadvantage of having to use modest instruments. The "personal equation" could not be ignored. Among the components contributing to this, the observer's eyesight was obviously of prime importance. As Percival Lowell wrote: "The eye is the portal to perception; through it is determined what shall enter the brain."[26]

Barnard, with his detection of the fifth satellite of Jupiter and his ability to measure accurately the positions of the satellites of Mars when someone with average vision "despaired of ever being able to set a wire upon them," had proved beyond doubt that he had an eye that was extraordinarily effective in detecting objects of extreme faintness.[27] Yet just as had been alleged in the debate between large and small telescopes—in which the former were granted superior power for detecting faint objects, owing to their light grasp, and the latter were considered superior when it came to fine planetary definition—there were supposed to be two correspondingly opposed qualities of sight. As Lowell put it again, "There are two ways in which eyes may admit information; by being sensitive to light or by being acute to form. . . . It is commonly assumed that because a man can see faint stars he must necessarily be able to detect planetary detail. There could be no greater mistake."[28]

The identity of the man who was able to see faint stars could hardly be missed. The very "sensitivity" of Barnard's eye was taken—by those for whom it was advantageous to so take it—as disqualifying it at the same time for planetary work, for the two qualities were supposed not to go together. "Just as it is paralleled in its capabilities by the large telescope," Lowell claimed, an eye like Barnard's "is peculiarly at home with a big instrument. The greater the illumination the better it is pleased."[29] The possessor of the "acute eye," on the other hand, would be better served by the smaller telescope, for its excellence was presumed to lie in the detection of planetary detail.

As was characteristic of him, Lowell reiterated the argument in essentially the same terms over the years. Of all the statements of his position, the following is, perhaps, the most developed:

> The divorce of the two qualities seems to be due to inherent incompatibility of kind. . . . The inner surface of the retina of the eye is a mosaic of cones and rods except in the central or yellow spot where only cones are present. This spot called also the fovea, is the point of distinct vision. The image falls upon the top of the cones or rods and the disturbance there produced is thence transmitted by nerve fibres to the optic nerve to be recorded in the

brain. Each cone transmits a point of the image. According, therefore, to the size of the cones will the impression be particularized; and that for the following reason. If the waves from two portions of an object fall upon one and the same cone its fibre will simply be more strongly stirred . . . but it will still produce but a single sensation in consequence of its solitary nerve fibre. To cause two separable perceptions two cones must be struck. Thus the diameter of the cones determines the distinction between the parts of an object. . . . The finer the cones the more delicate the perception. To have the foveal cones fine constitutes what we call an acute eye.

On the contrary the coarser the cones the greater the volume of sensation, we may suppose, their nerve fibres can carry to the brain. They thus send a more insistent message for the same outside call or in other words perceive light better, though they less minutely report the words of the dispatch. The message they convey is cheerful but inarticulate, as Stevenson happily puts it about another one. It thus appears not only that the two kinds of eyes are essentially different but are mutually exclusive.[30]

Of course, Barnard's sensitive eye, for all this inability to make out the spots of Saturn, had made out that embarrassing plethora of details in the Martian "seas" in 1894. Yet Lowell had a way of explaining that away as well. The "jumble of markings impossible to decipher," which was Lowell's paraphrase of what Barnard had seen, was but the "confused and blurred" form under which the true features of the planet—the network of fine lines—appeared to his sensitive eye.[31] An acute eye like Schiaparelli's (or, it is hardly necessary to add, Lowell's own) was needed to see them as they really were.

We shall find later that when other observers—noted from the first for their planetary work rather than for their detection of faint stars—also came to see the Martian "seas" as a "jumble of markings," a few years later, Lowell came up with equally ingenious arguments to explain these away, dependent this time not on differences in eyesight but on subtle atmospheric considerations.

Lowell was not the first to make the distinction between the sensitive and the acute eye. Something like it had been hinted at long before by the Jesuit astronomer F. De Vico, whose observations of Venus were discussed earlier. De Vico noted that in these studies the observers who were the most successful in seeing the faint markings on the planet were also the most unsuccessful in seeing the minute companions of bright stars. Rev. T. W. Webb, who cites De Vico's results in his *Celestial Objects for Common Telescopes,* finds the explanation "obvious enough, and worthy of notice." "A very sensitive eye," he wrote, "which would detect the spots more readily, would be more easily overpowered by the light of a brilliant star, so as to miss a very minute one in its neighborhood."[32]

However, Webb failed to explain why the sensitive eye would not also be overpowered by the glare of the planet, especially that of Venus, which in the telescope appears so dazzling to the eye.

Webb's friend, Rev. T. E. Espin, commenting on Webb's own keenness of sight for planetary detail in a reminiscence attached to the 1917 edition of Webb's *Celestial Objects,* recalls a striking instance of difference of vision between Webb and himself:

> A curious instance of difference of vision was well illustrated one superb evening, when Mr. Webb and the writer were observing Saturn with the 9⅓-in. reflector at Hardwick. Mr. Webb saw distinctly the division in the outer ring [Encke's] which the writer could not see a trace of, while the writer picked up a faint point of light, which afterwards turned out to be Enceladus, which Mr. Webb could not see.[33]

That such cases occur there can be no doubt. But they can just as easily be explained by differences in training of the eye as by physical differences; indeed, it is known that two observers with different specializations can quite readily be trained to see detail in the other's area of concentration.[34]

Finally, though Lowell insisted that the two qualities of eyesight "did not go together"—citing Schiaparelli as an example of acute, and Barnard of sensitive, vision—in fact it is surprisingly easy to find counterexamples. William Noble, who had discussed the matter briefly in 1877, had pointed out that Dawes excelled in the detection of both faint stars and planetary details.[35] So did those reputed possessors of "marvelous vision" Ward and de Boë. The average eye is said to be able to see six or eight of the stars in the Pleiades. Denning, a planetary observer of undoubted skill, saw fourteen, and Lowell—who wanted to believe that he had an acute eye—made out no less than sixteen, which would certainly make him more of a sensitive than most![36] My own experiments of a few years ago are consistent with this in showing that those who had the most acute also tended to have the most sensitive eyes.[37]

One can hardly escape the conclusion that the raison d'être of the theory of the two types of vision had more to do with rebutting Barnard's formidable observations than anything else. Of those observations, E. M. Antoniadi later wrote that they were "a rebuke from which the spider's webs [of the canals] were never to recover."[38] But this is clearly an exaggeration. The fact of the matter is that they did recover; indeed, as we shall see, their heyday was yet to come.

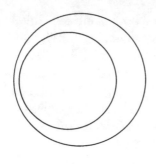

## CHAPTER 12

# "The Most Brilliant Man in Boston"

*Science would be far less advanced than she is if the passionate desires of individuals to get their own faiths confirmed had been kept out of the game.*
—*William James,* The Will to Believe

IN MARTIAN RESEARCH as we have traced it so far, the line of development has run more or less straight. As in the parlor game Twenty Questions, errors are gradually corrected, a more accurate concept of the planet gradually evolves. Suddenly one error—canals in the artifactual sense—is not corrected. It should have been with Barnard's observations of 1894. But that same year an individual of strong will and imagination, one willing and requiring the subordination of perception to imagination, seized on the Martian canals and, as it were, bent them to his own inner needs. Percival Lowell has been lurking about in these pages since the first chapter; it is time for him at last to claim center stage.

LOWELL WAS BORN in Boston on March 13, 1855, the eldest son of Augustus and Katharine Bigelow Lowell. There was blue blood on both sides of the family. An ancestor, Percival Lowle, had come to what later became known as Newbury, Massachusetts, in 1639. A Bristol merchant, he followed John Winthrop, the "Lodestone of America," to the Massachusetts Bay Colony at the advanced age of sixty-seven, and his bones lie in a little burial ground just north of the old town green in Newbury.

The Lowell type, through the intermarriage of cousins, was to remain remarkably stable for generations, going back at least to Rev. John Lowell of the Harvard class of 1721, as a series of portraits testifies. Moreover, the same abilities tend to recur. Among these the most pertinent for our purposes were in the mathematical and literary lines. Percival Lowell had these in unusual degree, as well as a personal magnetism that many who knew him noted. Thus journalist Ferris Greenslet once wrote: "This reporter has met many of the so-called great men of his time, but none with a more potent personal quality than Percival Lowell. He agrees with another witness that one felt it before, or almost before, he entered the room. It was as if one had been suddenly deposited in a powerful magnetic field."[1] But Lowell was far from universally admired, and in 1894 Harvard president Charles W. Eliot wrote the director of the Harvard College Observatory, E. C. Pickering, to quite the opposite effect: "Mr. Percival Lowell is undoubtedly an intensely egoistic and unreasonable person. . . . Fortunately he is generally regarded in Boston among his contemporaries as a man without good judgment. So strong was this feeling a few years ago that it was really impossible for him to live in Boston with any comfort."[2] No doubt the later furor over his Martian theories owed much to Lowell's strong personality. Though he did have his self-doubts, they were not usually apparent to others—certainly not to those outside the "clan"—and what appeared was too often rather a Brahmin sense of superiority and a tendency to dogmatism that, much as it may have assisted the popular reception of his views, did not always favorably impress his astronomical peers.

The early Lowells, though eminent in Massachusetts affairs, were not remarkably wealthy. This changed in 1813, when Francis Cabot Lowell, Percival's great-great uncle, on returning from England, where he had taken notes and made sketches of the machinery in the Lancashire textile mills, decided to build his own cotton mill at Waltham, Massachusetts, with spinning machinery and a practical power loom. "I well recollect," said Nathan Appleton, a leading stockholder, "the state of admiration and satisfaction with which we sat by the hour watching the beautiful movement of the new and wonderful machine, destined as it evidently was to change the character of all textile industry."[3] In the office of the textile business of his paternal grandfather, John Amory Lowell, Percival Lowell would later work and amass his own fortune. On the other side of the family, his maternal grandfather was Abbott Lawrence, minister to the Court of St. James's under Presidents Taylor and Buchanan—and indeed the Lawrences were as distinguished as the Lowells, with their own textile fortunes. Both had Massachusetts cities named for them.

Lowell's brother, Abbott Lawrence Lowell, would write that "the particular assortment of qualities a man inherits from among the miscellaneous lot his ancestors no doubt possessed and might have transmitted him, is of primary importance to him. In this Percival Lowell was fortunate."[4] But Percival Lowell's position was not altogether an enviable one, for part of his legacy was to have to live in the long shadow of a family that had accomplished much and expected more. As Lowell's brother testified, "Our father made us feel that every self-respecting man must work at something that is worthwhile, and do it very hard. In our case, it need not be remunerative, . . . but it must be of real significance."[5] Reputed to be a martinet with his millworkers, Augustus was presumably no less so with his sons.[6]

After being educated at various private schools at home and abroad, Percival Lowell attended Harvard, where he excelled in English composition and mathematics. He won the Bowdoin Prize for his essay "The Rank of England as a European Power from the Death of Elizabeth to the Death of Anne" and graduated with honors in 1876, giving as his part in the commencement exercises a talk on the nebular hypothesis of Laplace. His cousin, the poet James Russell Lowell, referred to him as "the most brilliant man in Boston."[7] One of his professors, the distinguished mathematician Benjamin Peirce, thought him "the most brilliant mathematician of all those who have come under my observation" and invited him to stay on at Harvard to teach the subject.[8] Lowell declined— "because I preferred not to tie myself down," he later recalled, "not because mathematics had not appealed to me as the thing most worthy of thought in the world."[9] Instead, with his cousin and freshman roommate Harcourt Amory, he opted for the Grand Tour of Europe as far as Syria, and though he had so recently refused to tie himself down to "the thing most

*Figure 12.1*
*Percival Lowell (1855 – 1916) as a*
*Harvard student.*

worthy of thought in the world," he almost managed to get himself enlisted for the front in the war between Serbia and Turkey!

On returning to the United States in 1877, the year of Schiaparelli's discovery of the canals, and still having no impulse to a profession, he went to work in his grandfather's textile business where he stayed for six years, "learning the ways of business, for a time acting as treasurer, that is, the executive head of a cotton mill," and managing trusts and electric companies. In 1883 he left the family business and set sail for the Far East, the first of three voyages there over the next decade. What prompted his decision to do so is not known with any certainty, but one might take a hint from Eliot's comment, quoted above, that Lowell's egoism made it "really impossible for him to live in Boston with any comfort." Alternatively, Ferris Greenslet speculates that "his *Wanderlust* may have been heightened by the impulse to evade cousinly caps that were being set at him." Greenslet cites in support of this the following remark from Lowell's book *The Soul of the Far East,* in which is compared the ancestor worship of the Japanese Buddhists and Boston Brahmins who "make themselves objectionable by preferring their immediate relatives to all less connected companions, and cling to their cousins so closely that affection often culminates in matrimony, nature's remonstrances notwithstanding."[10] The reader may judge which of these explanations, if either, is more plausible.

Evidence of anxiety surfaces from time to time in the record of these years. In 1884, on returning from his first Far Eastern trip, Lowell confided to his mother: "As for me, I wish I could believe a little more in myself. It is at all times the one thing needful."[11] Not long before, he had hesitated to accept a position as foreign secretary and general counselor to the embassy sent from Korea to the United States, "mainly due," as his brother later put it, "to anxiety to what his father would say."[12] Yet his self-doubts were apparently overcome; at least, as Ferris Greenslet remarks, "this was the last occasion in which there is any record of such distrust."[13] Indeed, the situation in Seoul proved an appealing one to someone with Lowell's poetic temperament, as his description of the house he was given there indicates: "From the street you enter a courtyard, then another, then a garden, and so on, wall after wall, until you have left the outside world far behind and are in a labyrinth of your own."[14]

LOWELL MADE the experiences of his maiden voyage to the Far East the basis of two books, *Chosön: The Land of the Morning Calm* (1886) and *The Soul of the Far East* (1888). The latter was to prove the most popular of his four Far Eastern books (*Noto* [1891] and *Occult Japan* [1895] were based on later voyages). In certain respects *The Soul of the Far East* is the key to Lowell's own soul more than any of his other writings, including his later, astronomical ones.

Just as the idea of natural selection—"survival of the fittest"—reflected the Victorian England of Malthus and laissez-faire, so Lowell's view of the Far East was conditioned by the Social Darwinism and industrialism of nineteenth-century New England. "After long groping about for some measure of orienting oneself," he asserted with a characteristic Lowellian pun, "one lights at last upon the clue. This clue consists of the 'survival of the unfittest.' "[15] The oriental peoples, Lowell contended, with the ethnocentrism typical of the age, respresented the "survival of the unfittest" because their evolution had stopped prematurely—or more precisely, he said, it had already run its full course. "Development ceased," he declared (making an astronomical allusion that foreshadows his later interest in the evolution of worlds),

> not because of outward obstruction, but from purely intrinsic inability to go on. The intellectual machine was not shattered; it simply ran down. To this fact the phenomenon owes its peculiar interest. For we behold here in the case of man the same spectacle that we see in the case of the moon, the spectacle of a world that has died of old age.[16]

The stagnation he found among Eastern peoples Lowell attributed to their "impersonalism"—their lack of individuality—which had impressed him after only two weeks in Japan. He had written to his mother: "Perhaps, a key to the Japanese is impersonalism. Forced upon one's notice first in their speech, it may be but the expression of character."[17] One can only add that the tendency of the newcomer to view a foreign people in such stereotyped terms probably suffices to explain the greater part of this impression.[18]

Attempting to make his abstract concept of impersonalism more concrete, Lowell described how individuality, its antithesis, first arises in childhood. "Very early in the course of every thoughtful childhood," he wrote, "an event takes place, by the side of which, to the child himself, all other events sink into insignificance." He continued, "Hitherto he has been aware only of matter; he now first realizes mind. Unwarned, unprepared, he is suddenly ushered before being, and stands awestruck in the presence of—himself."[19] Though this coming to self-awareness or acquiring the sense of one's individuality, Lowell assures us, must take place to some extent among all races of men, it does not do so to the same degree:

> It is one thing to the apathetic, fatalistic Turk, and quite another matter to the energetic, nervous American. Facts, fancies, faiths, all show how wide is the variance in feelings. With them no introspective γνῶθι θεαυόν overexcites the consciousness of self. But with us, as with those of old possessed of devils, it comes to startle and stays to distress. Too apt is it to prove an ever-present, undesirable double. . . . To this companionship, paradoxical though it sound, is principally due the peculiar loneliness of childhood. For nothing is so isolating as a persistent idea which one dares not confide.[20]

Here at least Lowell seems to be writing with some of the poignancy of self-revelation. What he is describing seems to be not so much the emergence of a sense of self per se as of a painful self-consciousness, in his own case early awakened, perhaps, under the watchful eye of an exacting and critical parent. Significantly enough, Lowell was to find the oriental peoples "still in that childish state of innocence before self-consciousness had spoiled the sweet simplicity of nature."[21] In poetic metaphor, of course, traveling east has always meant traveling back into the past—in the East lies the Old World and, in legend, Eden, the lost garden of innocence itself. At some deeper level of his unconscious psyche, Lowell seems all his life to have been setting out for a nostalgic world he once knew but later lost, "before self-consciousness had spoiled the sweet simplicity of nature," a world that would be a psychological haven for him and that was represented for a time by Korea, the "Land of the Morning Calm," and later by the world he was to discover in the planet Mars, of which he was to write, evoking Eden: "I may say that the translation onto Mars of the Phison and the Gihon, the two lost rivers of Mesopotamia, satisfactorily accounts for their not being found on Earth by modern explorers."[22]

LOWELL HAD BEEN interested in astronomy from boyhood, ever since as a youth of sixteen he had observed the heavens with a 2¼-inch refractor from the cupola on the roof of Sevenels, the family mansion on Heath Street in Boston. One finds in *The Soul of the Far East* evidence of his continuing fascination with other worlds and an idealistic strain as to what might be unfolded in them:

> In those bright particular stars—which the little girl thought pinholes in the dark canopy of the sky to let the glory beyond shine through—we are finding conditions of existence like yet unlike those we already know. . . . Conditions may exist there under which our wildest fancies may be commonplace facts. There may be
>
> > "Some Xanadu where Kublai can
> > A stately pleasure dome decree,"
>
> and carry out his conception to his own disillusionment, perhaps. For if the embodiment of a fancy, however complete, left nothing further to be wished, imagination would have no incentive to work.[23]

In *Purchas: His Pilgrimage* (1613), Xanadu is described as the place where Kublai Khan had built "a stately Palace, encompassing sixteen miles of plaine ground with a wall, wherein are fertile Meddowes, pleasant springs, delightfull Streames, and all sorts of beasts of chase and game, and in middest thereof a sumptuous house of pleasure."[24] Lowell knew of it from the poem that Purchas inspired, Coleridge's famous opium-dream poem, in which the lines Lowell intentionally misquotes appear. Whatever of Xanadu Lowell had expected to find in the Far East, there can be little doubt that by the time he made his last trip there, setting sail in December 1892, his conception of it had been pursued "to his own disillusionment." It was premonitory of the change in direction his career was presently to take that he carried along with him a six-inch Clark refractor, which he used to observe Saturn.

The planets, Lowell had intimated, might be places where "our wildest fancies may be commonplace facts." If the Far East had fallen short of the romantic ideal for which he had consciously or unconsciously yearned, he already had, by the time of his return to Boston in December 1893, a new world to explore. The New World discovered by Columbus became for generations of Europeans who set their bearings on America's shores—including, of course, Percival Lowell's own namesake eleven generations before—a reincarnation of the dream of the lost garden. Now, as the nineteenth century drew to its close, another Italian had navigated another ocean and had found another New World.

The comparison of Schiaparelli to Columbus was Lowell's own and is perhaps nowhere more eloquently stated than in the book he was to dedicate to Schiaparelli, *Mars and Its Canals* (1906):

> To Schiaparelli the republic of science owes a new and vast domain. His genius first detected those strange new markings on the Martian disk which have proved a portal to all that has since been seen. . . . He made there voyage after voyage, much as Columbus did on Earth, with even less of recognition from home.[25]

Yet Schiaparelli, also like Columbus, had failed to recognize the true import of his discovery. Thus, if Schiaparelli was Columbus, then Lowell was—or so he implied—a kind of Amerigo Vespucci of Mars. Yet in one sense he was Columbus too: he would travel west to find the new world in Mars by a route no one had ever taken before. The new route lay in that for which he headed west in the first place: better air. To get it he left behind those in observatories back east who toiled in vain under unfavorable skies. It was this, in large part, that would make it possible for him to make discoveries, or so he often insisted. "To see into the beyond requires purity," he wrote, "as little air as may be . . . and makes [one] perforce a hermit from his kind."[26] The East had been spoiled; in the pure and unsullied West, then, he would carry out his wish: to find the Xanadu in which to decree his "stately pleasure dome."

ACCORDING TO his long-time friend George Russell Agassiz, who accompanied him on the last of his three Far Eastern trips, Lowell decided to embark on his investigation of Mars upon hearing that Schiaparelli was being forced to retire from active planetary work owing to failing eyesight.[27] Considering the importance of Schiaparelli's work to the planetary astronomy of the period, anything that can be discovered about his eyesight is, of course, of the greatest interest. The gradual onset of his blindness is well known, and Lowell later described it in characteristically poetic terms:

> The eye which had peered so far beyond its contemporaries nature veiled with the coming of age. Perhaps the work it had been called upon to do had tired it before its time, perhaps its own delicate organization set bounds to its endurance; whatever the cause, in the early nineties it sank as it were gently to sleep. Slowly but steadily it waxed less, until for astronomic purposes it grew blind. He could still use it a little for everyday existence but celestially considered that wonderful eye had closed.[28]

Concerning Schiaparelli's visual powers at their peak, however, Robert G. Aitken, a highly respected observer of double stars at the Lick Observatory, said of his measures of the double star β Delphini that "the residuals shown in Schiaparelli's measures with the 8-inch telescope will not seem very large—in fact it is surprising that measures of such a pair could be obtained at all with so small a telescope."[29] Of the keenness of sight needed to make these measures, at least, there can be no doubt. Similarly, there was, and is, universal agreement as to the precision in his measurements of outlines of the dark areas on Mars. As for the canals, one can certainly agree with the comment of E. M. Antoniadi: "As a patient record of fleeting impressions, his results stand unrivalled."[30]

Rev. T. W. Webb mentioned on one occasion that Schiaparelli was "inconvenienced" by color blindness.[31] This is intriguing, as it seems possible that color blindness may have

obscured for Schiaparelli the blending of one shade into another on the Martian surface, which, as we have seen,
Nathaniel Green and others believed his hard, sharp lines
tended to misrepresent. Perhaps the precise outlines of planetary surface features that he recorded were thus a trick of the
eye. Schiaparelli had that kind of vision that might be called
"micrometric." Yet as E. Walter Maunder once observed, for
planetary work "what is required . . . is not so much keenness
of vision to detect minuteness of detail as power to appreciate
delicate differences of tone."[32] This Schiaparelli's eye seems to
have been less successful in discerning. Also, he tended to use
relatively high magnifying powers with his telescope, which,
as Maunder and Sir William H. M. Christie pointed out, also
"made him lose the half-tones of Mars."[33]

In any case, in light of the comment attributed to Lowell
that he took up Mars because he had learned of Schiaparelli's
failing eyesight, it is interesting that Schiaparelli, in 1892–93,
may not have known himself that his eyesight was failing, for
he later wrote to Antoniadi:

> As for my observations of Mars, alas! I have made none
> since 1899. I must admit that the *completely satisfactory*
> observations, made under good atmospheric conditions
> with the full power of my left eye (the other is imperfect
> in some respects) ended in 1890. During the oppositions
> of 1892-4-7-9 I still observed, but I made no drawings,
> because during this period no images were comparable
> with those of the period 1877-9-82-4-6-8. In 1890 I at-
> tributed this to the perpetual agitation of the atmosphere;
> the planet was low down, as was also the case in 1892.
> During 1894 (or 1896) a deterioration in my eyesight was
> becoming more and more evident; the field of vision be-
> coming darker and darker. At last, in 1900, I realized,
> with great sorrow, that the images were also becoming
> deformed. Then, I made my farewell. I also decided not to
> publish my observations of Mars made after 1890.[34]

Even if Schiaparelli realized that his eyesight was failing by
1894 or 1896, this would apparently not have been soon
enough to have occasioned Lowell's decision to take over for
the master, as the story goes. It may be that Lowell's decision
was based on others' views about Schiaparelli's eyesight rather

than Schiaparelli's own views, or perhaps Schiaparelli did not remember correctly. This was certainly the case in his telling Antoniadi that he had made no drawings in 1892, for one exists from September 15, 1892; it was published by Camille Flammarion in his *La Planète Mars et ses conditions d'habitabilité*. As for the detailed observations that had served as the basis of his series of memoirs on the planet, those of 1890 were indeed the last, and they were not published in their entirety until 1910, the year of his death, by which time he was totally blind.

The publication of Flammarion's *Le Planète Mars,* a massive review of all Martian observations through 1892, may itself have played an important role in directing Lowell's thoughts toward Mars in the period immediately before he established his observatory. Lowell received the book from an aunt shortly after his return from Japan and duly inscribed and dated it December 1893, also scrawling the word "Hurry" across the page—a not-so-cryptic entry, perhaps, in light of the fact that he would have equipped and operating a new observatory for the purpose of studying Mars by the end of the following May. According to Flammarion, Lowell himself later said that "his passion for astronomy, and in particular for the discoveries he made in the world of Mars, had been inspired . . . by the publication" of Flammarion's book.[35]

Flammarion had been born at Montigny-le-Roi in the department of Haute Marne in 1842. His parents hoped that he would oneday be ordained a priest. However, his interest in astronomy was aroused by an eclipse he witnessed at the age of five, and already by the time he was eleven, when he entered the ecclesiastical seminary of the Cathedral of Langres, his passion for astronomical observation was well developed. It was clear that if he consecrated himself to any priesthood, it would be a scientific one.

A child prodigy, by the time he was sixteen Flammarion had produced a manuscript of some five hundred pages (it was published in revised form in 1885 under the title *Le Monde avant la création de l'homme*—The World since the Creation of Man). A physician who had come to his home to treat him for an illness happened to notice the manuscript, asked to read it, and was sufficiently impressed to call it to the attention of the great Urbain Jean Joseph Leverrier himself, then director of the Paris Observatory and world famous for the calculations that had led to the discovery of Neptune in 1846. At sixteen Flammarion became junior assistant to Leverrier. Unfortunately, Leverrier was a rather disagreeable man. When Flammarion, at only nineteen, published *La Pluralité des mondes habités,* Leverrier summoned him into his office and said sourly: "I see, Monsieur, that you do not have to remain here. No, it is very simple. You can retire."[36]

In fact, there was some truth to Leverrier's comment, for this, Flammarion's first published book, made him a celebrity. In it he defended the thesis that the planets were inhabited. Thus he could truthfully write, in *La Planète Mars,* of having undertaken "even in our adolescence, in a scientific and literary career, . . . precisely the defense of the doctrine of a plurality of worlds,"[37] a doctrine that he was later to regard as positively proved with Schiaparelli's discovery of the canals. Indeed, long before Lowell entered the field, Flammarion was enthusiastically portraying Mars as a planet "very much alive." In *La Planète Mars* he had written: "It is the first time since the origin of mankind that we have discovered in the heavens a new world sufficiently like the Earth to awaken our sympathies."[38] Referring to the canals, which he was convinced were a system of waterways, he wrote that the great variations observed in them

> testify that this planet is the seat of an energetic vitality. These movements seem to us to take place silently because of the great distance separating us; but while we quietly observe there continents and seas . . . and wonder on which of these shores life would be most pleasant to live, there might at the same moment be thunderstorms, volcanoes, tempests, social upheavals, and all kinds of struggles for life. . . . Yet we may hope that, because the world Mars is older than ours, humankind there will be more advanced and wiser. No doubt it is the work and noise of peace that for centuries has animated this neighboring home.[39]

There are echoes of this in Lowell's writings.

Flammarion also believed, by the way, that there was vegetation on the Moon (an idea that was to be developed some years later by W. H. Pickering),[40] and like Father Secchi he argued that the whole universe must be teeming with life, writing: "We feel that unknown brothers are living there in other fatherlands of Infinity."[41]

Works such as *La Pluralité des mondes habités* and *La Planète Mars,* which Flammarion turned out with apparent effortlessness, were phenomenally successful. Among Flammarion's admirers was a very rich landowner from Bordeaux who was so moved by Flammarion's writings that he donated to him a magnificent château and park situated at Juvisy-sur-Orge, due south of Paris and lying between it and Fontainebleau. In the beautiful château, which had been built in 1730, Flammarion in 1883 set up a 9½-inch refractor. At about the same time he founded the Société Astronomique de France, an organization that encouraged amateur planetary observations. Among those who contributed to the society's bulletin was the Greek-born E. M. Antoniadi, who in the 1890s joined Flammarion at Juvisy and produced with him several canal-filled maps of Mars. Juvisy became something of a nerve center for Martian research. Percival Lowell visited Flammarion there every other year—during the off-years of Martian observation—and in a letter to his father he described dinner with Flammarion during one of these visits: "There were fourteen of us, and all that could sat in chairs of the zodiac, under a ceiling of pale blue sky, appropriately dotted with fleecy clouds, and indeed most prettily painted. Flammarion is nothing if not astronomical. His whole apartment, which is itself *au cinquième,* blossoms with such decoration."[42] In his correspondence with Flammarion, Lowell addressed him as *cher ami Martien* (reserving for Schiaparelli alone the loftier *cher maître Martien*).

Mars and extraterrestrial life were not the only exotic interests shared by Lowell and Flammarion. They were both fascinated by psychic research—as, for that matter, was the *maître Martien,* Schiaparelli himself, whose knowledge "that the framework of science is . . . neither firm nor fixed, and that we will only distantly understand the whole of the forces of nature," as Flammarion wrote, led him "to uncover the features that are inherent in the manifestations of psychic forces and the phenomena of spiritualism."[43] Other distinguished scientists of the period who were involved in such investigations included the biologist Arthur Russel Wallace, with Darwin codiscoverer of the principle of natural selection, and the physicists Sir Oliver Lodge and Sir William Crookes.

To the lay mind, of the same order as psychic research and spiritualism was hypnosis, then all the rage in intellectual circles. Charcot had done public demonstrations in Paris showing that by hypnotic suggestion he could produce in subjects hysterical paralyses and contractions having the same features as those produced in spontaneous attacks. These demonstrations made a deep impression on Freud, among others, who had gone to Paris specifically to study with Charcot and who came away with the conviction that "there could be powerful mental processes which nevertheless remained hidden from consciousness."[44] Lowell took up the fashionable subject with characteristic gusto, setting sail on his last Far Eastern voyage (the one on which he carried along the telescope) in order to study Shinto trances, even holding séances in his own house—"red flame and potent spells in a dark dark room," he wrote of them to a friend.[45] Among other things, he examined the possessed by taking their pulses and sticking pins into them to test their sensibility while in the trancelike state.[46]

Flammarion's interest in such exotic research was more profound and certainly more enduring, dating back to his early twenties and continuing right up to his death in 1925. In 1899 he began assembling thousands of documents from correspondents who claimed to have experienced premonitions, telepathic communications, and the like, which he believed to be transmitted by means of vibrations of the ether, "projected to a distance like all vibrations of ether," and felt by those brains in harmony with the "psychic force" producing them, the analogy being to the way the telephone worked (in those days still a novel invention):

> The transformation of a psychic action into an ethereal movement, and back, may be similar to that which we observe in the telephone, where the receiving disk, which is identical with the sending disk, recreates the sound-movement that was transported not by sound but by electricity.[47]

Though in later years such investigations came to absorb more and more of Flammarion's energies, he always regarded him-

self as an astronomer first and foremost, and apparently he saw
*la pluralité des existences de l'âme* as simply a natural corollary
to *la pluralité des mondes habités.*

AFTER SCHIAPARELLI and Flammarion, Lowell probably
owed his greatest debt to fellow Bostonian William H. Pickering of the Harvard College Observatory, who was the younger
brother of the observatory's director, Edward C. Pickering. It
was he who introduced Lowell to the importance of atmosphere in planetary observation.

For some years Harvard astronomers had been actively in-
vestigating various observing sites in order to find one with
superior atmospheric conditions. The search was prompted by
the bequest of Boston mechanical engineer Uriah A. Boyden,
who in 1887 had willed the observatory $238,000 for the
establishment of an observing station "at such an elevation as
to be free, as far as practicable, from the impediments to
accurate observation which occur in the observatories now
existing, owing to atmospheric influences."[48] The summer of
1888 saw E. C. and W. H. Pickering testing sites in Colorado,
setting up a 12-inch Clark refractor at Colorado Springs,
Leven Lakes (near Pikes Peak), and atop Pikes Peak itself
(elevation 14,147 feet). However, the steadiness of the seeing
did not seem to surpass that at Cambridge, and they concluded: "The selection of a proper site for an observatory is by
no means merely a question of elevation."[49]

W. H. Pickering went farther west, to California, and that
winter he began a year of tests there. Among the sites he
examined was Mt. Wilson, where he found that "the definition
was extraordinarily fine," but it was not then possible to secure
a deed to the land near the peak. Meanwhile, two other Harvard astronomers, S. I. and M. H. Bailey, were carrying out
tests in the Peruvian Andes. Arequipa, at an elevation of 8,100
feet, appeared particularly favorable. It was nearly 4,000 feet
higher than the only permanent mountain observatory established to that time, the Lick, at 4,200 feet on Mt. Hamilton.
In 1890 a permanent observing station was established on the
site under W. H. Pickering's direction. Andrew Ellicott Douglass, a recent graduate of Connecticut's Trinity College, was
his assistant.[50]

Pickering, observing with a 13-inch Clark refractor, found

seeing conditions at Arequipa superb during his brief sojourn there, and he reported quite sensationally:

Ten or twelve diffraction rings have been counted, under favorable circumstances, around brighter stars, each ring being nearly if not absolutely motionless. It is well known that in general to see the rings at all with a telescope as large as the 13-inch is a rare occurrence, and that the few there seen are nearly always wavering and broken.

At first a power of 475 was used extensively for all observations. . . . Since then higher powers have been sent from home, and 1140 diameters have been used on Venus in the daytime. . . . The phases of Jupiter's satellites are readily observed as they enter into the shadow of the planet, a phenomenon which it is thought but few astronomers have ever seen, even with much larger telescopes.[51]

The ability to distinguish the diffraction rings around bright stars was, by the way, to serve as the basis of a scale Pickering worked out for quantitatively evaluating atmospheric conditions at different sites, a scale later used at the Lowell Observatory (where, as the observing notebooks testify, seeing was usually about 6 on Pickering's scale of 10).[52]

With Mars high in the Peruvian sky in 1892, Pickering turned his telescope toward that planet as well. He sent back his results in a series of telegrams to the *New York Herald,* which carried reports of his observations, such as: "Mars has two mountain ranges near the south pole. Melted snow has collected between them before flowing northward. In the equatorial mountain range, to the north of the gray regions, snow fell on the two summits on August 5 and melted on August 7." Or again: "Prof. Pickering . . . says that he has discovered forty small lakes in Mars."[53] E. S. Holden at the Lick Observatory complained that "the essence of the scientific mind is conscientious caution; and this is especially necessary in referring to matters in which the whole intelligent world is interested—as the condition of the planet Mars. . . . How does [Pickering] know the dark markings are lakes? Why does he not simply call them dark spots?"[54] Pickering's brother, who perhaps knew best how to take these reports, wired back that the "telegrams to the New York Herald have given you a colossal reputation. A flood of cuttings have appeared, forty nine coming this morning." However, he added, "in my own case I should have restricted myself more distinctly to the facts. . . . You would have rendered yourself less liable to criticism if you had stated that your interpretations were probable instead of implying that they were certain."[55]

Pickering was relieved of the directorship of the observatory at Arequipa in 1893 owing to poor management of funds—and to the fact that he paid more attention to the Moon and planets than to the stellar observations that were supposed to be his chief business. His replacement, S. I. Bailey, wrote the following year to E. C. Pickering that he had no intention of spending much time observing Mars. "I fear," he admitted with perhaps more cynicism than modesty, "that I have not the creative faculty sufficiently developed to make a mark as an observer of Mars."[56]

Meanwhile, W. H. Pickering and Douglass, having returned by steamer from Peru to Boston, were introduced to Percival Lowell in January 1894—probably through the agency of Lowell's cousin, A. L. Rotch, who at one time had accompanied Pickering on an eclipse expedition to Chile. Though he had only recently returned from the Far East, Lowell had clearly been thinking for quite some time of astronomical themes. Already in July 1891 he had written to his brother-in-law William Lowell Putnam of a plan for a "philosophy of the cosmos, with illustrations from celestial mechanics."[57] Though he gave no details, the purport of this seems clear enough: in that late Victorian age, which had so recently witnessed the great achievements of Darwin, Wallace, Huxley, and Spencer, the "philosophy of the cosmos" could hardly have been formulated on other than evolutionary principles. The plan, in some form, was gradually to find expression in Lowell's writings over the next twenty-two years. The question of life on Mars fit for him within this wider framework, for within the scheme of "cosmic evolution" life was seen as an "inevitable detail." "Each body," he wrote, "under the same laws, conditioned only by size and position, *inevitably* evolves upon itself organic forms" (italics mine).[58]

Schiaparelli had apparently shown that conditions on Mars did not seem altogether antagonistic to life, and he had revealed a set of markings with singular implications, given

Lowell's premise that "life is an inevitable detail of cosmic evolution." Against this background, Lowell's meeting with Pickering was fateful, for Lowell became convinced at once that through his experiments at Arequipa, Pickering had "brought into cooperation a practically new instrument" for planetary research: the air itself.

There were thus several things that came together in Lowell's mind in late 1893, and it was nothing so simple as learning of Schiaparelli's failing eyesight that led him to conceive of the plan of establishing a new observatory. Its objectives were to be: "1st, the determination of the physical conditions of the planets of our Solar System, primarily Mars; 2d, the determination of the conditions conducive to the best astronomical observations."[59] Pickering and Douglass, who were granted leaves of absence of one year from Harvard, agreed to assist him in the project.

Indeed, the observatory was initially planned as a joint venture with the Harvard College Observatory. Not only did Lowell borrow its astronomers, he borrowed one of its telescopes, a 12-inch refractor. Apparently the understanding at Harvard was that the well-to-do Lowell proposed to serve primarily in the role of a patron, but Lowell soon made it clear that he had no intention of merely "going along," as the press reported, but wanted a full share of the recognition. Before long his official association with Harvard was somewhat acrimoniously dissolved.

Meanwhile, the site for the observatory had yet to be chosen, and this was a matter of some urgency, since Mars would come to opposition in October in what would be, in the cycle of oppositions, the last reasonably good opportunity to observe the planet for a number of years. To this purpose, Douglass was sent west in early March to scout for sites, taking along the 6-inch refractor Lowell had used in Japan (it is now mounted, together with the 12-inch telescope Lowell borrowed from Harvard in 1894, alongside the famous 24-inch in Flagstaff). Arriving in Tombstone by March 8, Douglass tested seeing there using Pickering's scale, then headed on to Tucson, Tempe, and Phoenix in southern Arizona before veering north to Prescott and Ash Fork. Finally he came to Flagstaff. The altitude at this site, 7,000 feet on the Coconino Plateau, appealed to Lowell, who wrote Douglass that "other things being equal, the higher we can get the better."[60] Douglass obtained good results in the few observations he made in the "opening in the woods" on the mesa just west of town, and on this basis Lowell decided on Flagstaff. Ironically, Douglass later came to the conclusion that "the San Francisco Peaks just to the north project into the stream of air moving overhead to produce eddies" detrimental to the best seeing.[61] Considering how supremely important Lowell felt that seeing was for planetary observation, it is, in fact, a little surprising that such a momentous decision should have been made on the basis of only eleven days of tests. Lowell himself later came to have serious private doubts about conditions at Flagstaff, leading to a rather ill-fated attempt to move the observatory to Mexico in 1896. Others have also expressed reservations. Clyde Tombaugh, who observed the planets often there as an assistant astronomer at the Lowell Observatory, has stated that "most of the time the seeing at Flagstaff is mediocre." Nevertheless, he did experience superb seeing at times, usually at twilight—"when the Sun's disturbing effect in producing convection turbulence is cut off, and before atmospheric inversion sets in"—or when there was "either no wind, or a westerly wind, and a falling barometer."[62] It does seem safe to conclude that Lowell tended to exaggerate the superiority of the seeing at Flagstaff, yet so insistent was he in his propaganda on the point that even his severest critics generally conceded him this advantage.

The decisiveness with which Lowell made up his mind about where to put his observatory points out, in any case, a striking characteristic of the man. As before, when he had within two weeks hit upon the "soul of the Far East," the great insights of his life were usually reached with lightning speed, as was only consistent, perhaps, with the family motto *Occasionem cognosce* (loosely, seize your opportunity). His major conclusions about Mars, which he would presently be arriving in Arizona to study, were similarly reached within two months, at most, of his first observations of the planet—if not, as critics alleged, even before his arrival—and he was to defend these conclusions with dogged persistence until his death twenty-two years later.

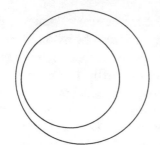

## CHAPTER 13

# The Visions of Sir Percival

*O pardon! since a crooked figure may*
*Attest in little place a million:*
*And let us, ciphers to this great*
*    accompt,*
*On your imaginary forces work.*
        —Shakespeare, Henry V

IN HIS EARLY writings on Mars, Schiaparelli had cautioned against taking the nomenclature of the maps too literally; later he acquiesced to an aquatic Mars. His own views about the planet in 1893, at the time Lowell was about to enter the field, were summarized in a paper later translated into English by W. H. Pickering.[1] The paper begins with a careful discussion of the polar caps, which, Schiaparelli says, must represent on Mars as on Earth an "immense mass of ice and snow." He points out that the conclusion follows not only from their analogous aspect and place on the planet but also from another important observation:

> It is manifest, that if the . . . white polar spots of Mars represent snow and ice, they should continue to decrease in size with the approach of summer in those places, and increase during the winter. . . . These observations of the alternate increase and decrease of the polar snows are easily made, even with telescopes of moderate power, but they become much more interesting and instructive when we can follow assiduously the changes in their more minute particulars, using larger instruments. The snowy regions are then seen to be successively notched at their edges; black holes and huge fissures are formed in their interiors; great isolated pieces many miles in extent stand out from the principal mass, and dissolving, disappear a little later. In short, the same divisions and movements of

these icy fields present themselves to us, at a glance, that occur during the summer of our own arctic regions, according to the descriptions of explorers.[2]

From these observations of the polar caps, Schiaparelli comes to several conclusions. The first is that the planet must be surrounded by an appreciable atmosphere, for the renewed snows must be the result of the precipitation of vapor carried by atmospheric currents and condensed by cold. Moreover, alluding to the pioneering spectroscopic observations of Sir William Huggins in England, Jules Janssen in France, and Hermann Vogel in Germany during the 1860s and 1870s, which had indicated that this vapor was in all probability water, he goes on to say:

> When this conclusion is assured beyond all doubt, another one may be derived from it, of not less importance,—that the temperature of the [Martian] climate, notwithstanding the greater distance of that planet from the Sun, is of the same order as the temperature of the terrestrial one. Because, if it were true, as has been supposed by some investigators, that the temperature of Mars was on the average very low (from 50° to 60° below zero!) it would not be possible for water vapor to be . . . an important factor in its physical changes; but would give place to carbonic acid, or to some other liquid whose freezing point was much lower.[3]

The idea that the polar caps might be frozen carbon dioxide (dry ice) was proposed by Ireland's G. Johnstone Stoney, who had used the molecular theory of gases recently formulated by James Clerk Maxwell to calculate that Mars was not massive enough to have hung onto much water.[4] Stoney was, by and large, right: we now know that the southern cap, which can entirely disappear, is wholly $CO_2$, while the northern cap as it shrinks leaves behind a mere remnant of water ice. But Stoney's work was given less credit than it deserved. The spectroscopic detection of water vapor in the Martian atmosphere was more difficult than appeared at the time, and Schiaparelli had excessive confidence in the early spectroscopic results. The method used was to compare the spectrum of Mars with that of the airless Moon to see if the absorption lines due to water vapor were intensified in the Martian spectrum. Unfortunately, the two bodies usually had to be observed in different positions in the sky, making it difficult to obtain spectra under identical conditions. The atmosphere of the Earth, so frustrating to visual observers, was, if possible, even more frustrating to pioneer spectroscopists. The moisture-laden blanket of terrestrial air superimposed lines of its own on those due to the planet's atmosphere, making the spectrum like a painted-over canvas. Finally, the prejudice of the observer was bound to enter into the assessment of the relative intensity of the lines. The conclusion reached by Huggins, Janssen, and Vogel, which Schiaparelli had hoped would be "assured beyond doubt," was within a year of Schiaparelli's writing challenged by W. W. Campbell, who used an improved spectroscope with the Lick 36-inch refractor in dry summer air over Mt. Hamilton. "The spectrum of Mars," Campbell wrote, "has appeared to be identical with that of the Moon in every respect."[5]

In all fairness, there was one other bit of evidence offered in addition to the spectroscopic observations to support the idea that liquid water was present on the Martian surface. Observers beginning with Beer and Mädler had noted that during the time of the shrinkage of the polar caps there appeared surrounding them a dark zone. The German astronomers, as we have seen, had thought these "marshes," and Schiaparelli regarded them as "a species of temporary sea." The bluish tint they appeared to have, whether due to chromatic aberration produced by the refractors in use in those days or to eye fatigue owing to contrast with the predominant ochre tinge of the disk, was in either case not that of liquid water, but few looked critically at the uncautious reasoning from analogy that underlay the assumption that it was. The point was a crucial one, for dry ice, at the atmospheric pressure of Earth—not to mention that of Mars, where it is far lower—sublimes, that is, passes directly from the solid to the gaseous state without becoming liquid at all. If there was liquid on the surface, as the blue borders of the caps were taken to prove, there could be little doubt that it was water. This line of reasoning, advanced by Percival Lowell, was later criticized by Alfred Russel Wallace, who called it "very weak and inconclusive," even "frivolous," and added:

It is perfectly well known that although water, in large masses and by transmitted light, is of a blue colour, yet shallow water by reflected light is not so; and in the case of the liquid produced by the snow-caps of Mars, which the whole conditions of the planet show must be shallow, and also be more or less turbid, it cannot possibly be the cause of the "deep blue" tint said to result from the melting of the snow.[6]

Returning now to Schiaparelli's 1893 paper, during the melting of the polar caps the canals also underwent marked changes, according to the Italian astronomer:

At that time the canals of the surrounding region become blacker and wider, increasing to the point of converting, at a certain time, all of the yellow region comprised between the edge of the snow and the parallel of 60° north latitude, into numerous islands of small extent. Such a state of things does not cease, until the snow, reduced to its minimum area, ceases to melt. Then the breadth of the canals diminishes, the temporary sea disappears, and the yellow region again returns to its former area. . . . The most natural and the most simple interpretation is . . . of a great inundation produced by the melting of the snows,—it is entirely logical, and is sustained by evident analogy with terrestrial phenomena. We conclude that the canals are such in fact, and not only in name.[7]

Yet while here admitting that the canals seemed in fact to form a "true hydrographic system," Schiaparelli still fell short of attributing to them an artificial character:

The network formed by these was probably determined in its origin in the geological state of the planet, and has come to be slowly elaborated in the course of centuries. It is not necessary to suppose them the work of intelligent beings, and notwithstanding the almost geometrical appearance of all of their system, we are now inclined to believe them to be produced by the evolution of the planet, just as on the Earth we have the English Channel and the Channel of Mozambique.[8]

It was Schiaparelli's 1893 paper that largely formed Lowell's own concept of Martian conditions and served as the framework of his own theory, which was developed in 1894 at least as much from Schiaparelli's summary as from Lowell's own observations. If the apocryphal story of Lowell's taking over Mars from Schiaparelli when he learned that the Italian observer's eyesight was failing cannot be fully credited, it captures, like most such stories, more than a modicum of truth, for in a manner of speaking Lowell was to take over Schiaparelli's Mars, viewing the planet through the categories that the latter had established, with the canals seen as a hydrographic system and, at first, the dark areas as seas, though that, at least, was to be rejected, and with its rejection the stage was set for Lowell's own interpretation of the canals' significance.

LOWELL DOCUMENTED his arrival at Flagstaff, on May 28, 1894, in a letter to his mother, to whom he was writing nearly every day: "Here on the day. Telescope ready for use tonight for its Arizonian virgin view. . . . Today has been cloudy but now shows signs of a beautiful night and so, not to bed, but to post and then to gaze."[9] In fact he was planning to use not one but two telescopes on Mars—the 12-inch he had borrowed from Harvard and an 18-inch Brashear refractor obtained from the University of Pennsylvania which, after its brief detour to Arizona, was destined for Philadelphia's Flower Observatory.[10] That night clouds moved in and rain fell through the still uncanvassed dome on Lowell and Pickering, who had gone there to camp to be ready for "early rising Mars."

On May 31 the sky finally cooperated and the 12-inch was used for a first view of the planet. The following night Lowell recorded in his notebook his initial impressions with the 18-inch: "Southern Sea at end first and Hourglass Sea [Syrtis Major] . . . about equally intense. . . . Terminator shaded, limb sharp and mist-covered forked-bay vanishes like river in desert." Lowell's use of the word *desert*, though perhaps only meant as a figure of speech, is nevertheless remarkable and suggests that the kernel of his later theory about the planet was already there. On June 2 he noted: "None of the dark parts . . . really darker than a grey."

He and Pickering recorded the Lethes on June 7, and he
gloated: "This is the first canal seen here this opposition, and
in all likelihood the first seen anywhere." Two days later he
found it "very broad and glimpsed double," but doubted himself: "These sudden revelation peeps may or may not be the
truth." On June 19 he wrote: "With the best will in the world I
can certainly see no canals." But in imagination he had already
gone far beyond what one might expect from the "glimpses"
and "suspicions" he logged in his notebook.

Lowell's thoughts about Mars at the time he began to observe it are documented in a hitherto unpublished poem entitled simply "Mars," which exists in the Lowell Observatory
archives and appears to have been written shortly after his
arrival. Though his secretary, Wrexie Louise Leonard, went a
bit far when she wrote that "in poetry he at times was touched
with divine power and was capable of producing sparkling
gems of which even great poets might be proud to say, 'This is
mine own,'" "Mars" is, in its way, deeply evocative, and it
conveys a good deal of the emotion with which Lowell that
summer embarked upon his Parsifalian quest.[11]

Lowell begins, fittingly enough, with an invocation comparing the quest for the world in Mars with a voyage across an
ocean:

> One voyage there is I fain would take
> While yet a man in mortal make;
> Voyage beyond the compassed bound
> Of our own Earth's returning round.
>
> .   .   .   .   .   .   .   .   .   .   .   .   .
>
> My far-off goal seems strangely near,
> Luring imagination on,
> Beckoning body to be gone
> To ruddy-earthed, blue-oceaned Mars.

The last phrase betrays the date of composition, in showing
that at this time Lowell still accepted the Schiaparellian Mars
of the Italian astronomer's 1893 paper. Publicly he maintained
that the "blue" (or, as they appeared to his eye, "grey-green")
areas on the planet were seas until July 1894, when several new
findings led him to reject the idea, though before that time, as
his notebooks show, he had grown privately skeptical.

In his poem Lowell concludes by saying that the lands and seas on Mars are "intertwined more happily" than on Earth. The seas there are "all Mediterranean"; the lands, lying "content," are wrapped by "bluish arms of the winding deep," to make Mars "a very Venice of a place." His reference to the Mediterranean may in part have been unconsciously inspired by the classical nomenclature of Schiaparelli's map. But the arms of the winding deep that enwrap the continents on Mars are for Lowell even here the retreating arms of vanishing oceans—the planet, "being smaller, aged more fast than the Earth," and it shows in its topography the "tell-tale traits of an older past":

> Surface through which half the oceans have sunk,
> Their once broad bosoms already shrunk
> To gourd-shaped straits whose gaining strand
> Foretells the time now close at hand
> When the last vestiges of seas
> Shall be swallowed. . . .
> And Mars like our moon through space shall roll,
> One waterless waste from pole to pole,
> A planet corpse, whence has sped the soul.

Lowell's theories of advancing planetary desertism were already, apparently, firmly in place, providing a ready-made interpretive framework for the observations before he can be said, properly speaking, to have put eye to telescope at all. The Schiaparellian nomenclature had exerted its claim on him, and the markings were unwittingly assumed to be what their names said they were, whether sea, bay, isthmus, or canal ("a very Venice of a place"). But the emphasis on desertism was his own unique contribution; and whatever the basis in Lowell's complex psyche, rather than continuing to project on to the planet the images of paradise that his poem had first invoked, a different, darker obsession soon asserted itself and colored the planet with apocalyptic images of doom.

On several occasions Lowell underscored the observation he had first made on June 2: how faint the dark areas appeared. On June 19, for instance, he wrote: "Markings on Mars always faint, so faint at times it seems impossible the dark ones should be seas."[12] This ought to have been a clue that perhaps the implied analogy was not a very exact one, that Mars, rather than being, or ever having been, another Earth, was in reality something quite different—"itself alone," with its own categories, of which the terms *continent* and *ocean* might not be adequate paraphrases. But the persistent image controlled the thought here as well: though no longer seas, they became the ghosts of seas in Lowell's mind—the dried-up basins of former oceans—and having begun by calling the salmon-colored continental areas deserts (perhaps at first as a mere figure of speech), he could now conclude that the planet was in fact on the way to utter desiccation. When his assistants—Pickering and Douglass—sketched canals in the dark areas, and finally when Pickering in July attempted to measure the polarization of the light reflected from them and found it inconsistent with a water-covered surface, Lowell regarded the point as cinched. The only water, then, was apparently that locked up in the polar caps and released when they melted each summer. This was the concept of Mars that took shape in his mind that first summer, and it was no doubt well along when he left Flagstaff for Boston in July. He returned to Flagstaff again in August for a second stint at the telescope. In justice to him it must be admitted that his image of an extremely arid planet was actually closer to the truth than the aquatic Mars, with its "winding deeps" and "inundations."

Whether interpreted as shallow seas or dried-up basins, the peculiar physiognomy of the Martian surface, whose continents everywhere interlaced with arms of the sea (or the basins of what had *once* been arms), had already suggested to earlier astronomers that there was probably little difference in the elevation of these areas. Indeed, Lowell—completely oblivious to the observations that E. E. Barnard was then making that the "seas" appeared "broken by canyon and slope and ridge," obvious indications of a marked unevenness of the surface—proceeded to explain away observations made at his own observatory that seemed, on the face of them, to suggest anything but a flat Mars. Seen at quadrature, when the phase of the planet is greatest, he wrote that "the terminator, or inner edge, presents a very different appearance from the lunar one. Instead of looking like a saw, it looks comparatively smooth, like a knife."[13] But Pickering and Douglass had made out a great number of irregularities of the terminator (some had also been observed at the two previous oppositions at Lick).

Lowell's argument was hard to follow on this point, but he showed, or seemed to show, that all of the irregularities could be dismissed as being due to clouds and that none of them, therefore, had a mountainous basis. W. W. Campbell, the Lick Observatory astronomer who had obtained the negative spectroscopic result concerning water vapor in the Martian atmosphere, was among those who found Lowell's argument unconvincing. "There are some arguments in favor of [the irregularities] being clouds," he admitted,

> but many more in favor of their being mountains. The observed phenomena are fully explained by supposing a mountain chain to lie across the terminator and to disappear from sight by the planet's diurnal rotation. The observed projections were such as would be produced by the Sun shining on the mountain tops outside the terminator, and the observed depressions were such as would be formed by the shadow of the mountain range lying within the terminator. . . . I confess my inability to unravel Mr. Lowell's discussion of Mr. Douglass's observations. When it was a question of detecting a twilight effect, it was the illuminated atmosphere which formed the visible terminator. When it was a question of proving that Mars was extremely level, and would therefore lend itself to general irrigation, it was the land surface that formed the visible terminator.[14]

Schiaparelli, by the way, tended to side with Campbell, noting that at least certain of the irregularities of the terminator could only be explained "by the presence of elevations or depressions on the surface of Mars."[15]

Later, Lowell's arguments for the supposed flatness of Mars were to turn flatly specious.[16] But in fact the point was of some importance, for a flat Mars was actually a prerequisite for his theory of intelligent life on the planet. Schiaparelli's canals followed straight lines (or their equivalents on a globe, arcs of circles). Lowell pointed out that if there were inhabitants of the planet, they would, owing to its dry condition, undoubtedly have to irrigate. Conveniently, there was, as he wrote, "a network of markings covering the disk precisely counterparting what a system of irrigation would look like;

and . . . a set of spots placed where we should expect to find the lands thus artificially fertilized"[17]—the "lakes" Pickering had reported from Arequipa now becoming transformed into "oases." Meanwhile, the dark areas of the planet, from having been seas, now became, according to Lowell, vast tracts of vegetation. This, in a nutshell, was what Lowell called his theory. However, for the canal theory to hold water, as it were, it was quite necessary for Mars to be relatively smooth, as he himself was well aware: "For the lines to contain canals we must suppose either that mountains prove no obstacles to the Martians, or else that there are practically no mountains on Mars. For the system seems sublimely superior to possible obstructions in this way."[18]

LOWELL PROCEEDED to unleash his theory on the world in a flurry of articles published in *Astronomy and Astro-Physics,* the *Atlantic Monthly,* and *Popular Astronomy*; in a series of well-attended lectures delivered at Huntington Hall in Boston in February 1895; and above all in a book, *Mars,* published in December 1895 as he set sail for Europe to confer with the likes of Schiaparelli and Flammarion before heading on to the Sahara to investigate the seeing conditions there.

The contrast could hardly be greater between Lowell, who as the veteran of only a single season of planetary observation could hardly be regarded as other than a relative tyro, and the more conservative Barnard, who did not publish his far more important observations for two years and then did so in fragmentary form. The fear of ridicule was apparently unknown to Lowell, at least during this "manic" phase of his career. With his book *Mars* he had become one of the most prominent men in Boston, to judge from a vignette from about this time related by Ferris Greenslet:

He had bought for his life *en garçon* a small high house on the upper side of West Cedar Street. There during the winter a young editor and publisher from New York, passing with a bag of manuscripts to his own modest establishment in the next block, used to observe him every weekday at five-thirty. His handsome head was to be seen *vis-à-vis* the *Boston Evening Transcript* beneath a life-sized plaster Venus similar to those that infest the Athenaeum. Visibility was perfect, for the shade was always raised to the very top of the window as if to admit no impediment to a message from Mars.[19]

Lowell's arguments for Martian life were from the first embedded in the matrix of his broad evolutionary scheme of planetary development. In *Mars* he wrote:

> A planet may in a very real sense be said to have life of its own, of which what we call life may or may not be a subsequent detail. It is born, has its fiery youth, sobers into middle age, and just before this happens brings forth, if it be going to do so at all, the creatures on its surface which are, in a sense, its offspring. The speed with which it runs through its gamut of change prior to production depends upon its size; for the smaller the body the quicker it cools. . . . Now, in the special case of Mars, we have before us the spectacle of a world relatively well on in years, a world much older than the Earth. To so much about his age Mars bears evidence on his face. He shows unmistakable signs of being old. Advancing planetary years have left their mark legible there. His continents are all smoothed down; his oceans have all dried up.[20]

Consequently, "any life he may support [should] be not only relatively, but really older than our own":

> From the little we can see, such appears to be the case. The evidence of handicraft, if such it be, points to a highly intelligent mind behind it. Irrigation, unscientifically conducted, would not give us such truly wonderful mathematical fitness in the several parts to the whole as we there behold. A mind of no mean order would seem to have presided over the system we see,—a mind certainly of considerably more comprehensiveness than that which presides over the various departments of our own public works. . . . Quite possibly, such Martian folk are possessed of inventions of which we have not dreamed, and with them electrophones and kinetoscopes are things of a bygone past, preserved with veneration in museums as relics of the clumsy contrivances of the simple child-

hood of the race. Certainly what we see hints at the existence of beings that are in advance of, not behind us, in
the journey of life.[21]

Lowell expected that some would be reluctant "to admit the
possibility of peers," but such reluctance as there was did not
come from the general public, which was, to the contrary,
more than willing to go along. Some idea of the way his
theory was received by the public can be gleaned from the
reception given his lectures. As Lowell's secretary described
one such affair, "Standing room was nil, and demands for
admission were so numerous and insistent that repetitions
were arranged for the evenings. At these repeated lectures the
streets near by were filled with motors and carriages as if it
were grand opera night!"[22]

Yet uncritical as was the reception by the lay public, some
members of the scientific community looked for and found
weaknesses in the chain of reasoning Lowell had seemingly
forged with such irresistible logic. W. W. Campbell, in a critical, but just, review of *Mars*, wrote:

> Let us examine Mr. Lowell's irrigation scheme. We are
> asked to believe that the equatorial region of Mars, form-
> ing a strip at least seventy degrees wide, can be and is
> irrigated from the north and south poles; the "canals" in
> the two cases of opposite flow being identical! The con-
> temporary problem on Earth would be to irrigate San
> Francisco, Chicago, New York, Rome, Tokyo from the
> snow melting at the South Pole; and to irrigate Val-
> paraiso, Cape of Good Hope, Australia, from the snow
> melting at our North Pole; all the irrigated land lying
> between New York, etc., on the north and the Cape of
> Good Hope, etc., on the south, to be irrigated alike from
> the North and South Poles.[23]

Campbell also did not fail to notice the extent to which Lowell's observations supported his "pre-observational views" as
stated in a lecture to the Boston Scientific Society delivered
two days before his arrival in Flagstaff. Lowell had said that
the "investigation into the condition of life on other worlds,
including last but not least, their habitability by beings like
[or] unlike man . . . is not the chimerical search some may

*Figure 13.1
Percival Lowell at the lecture
podium, delivering one of the
Lowell Institute lectures of 1906.*

suppose. On the contrary, there is strong reason to believe we are on the eve of pretty definite discovery in the matter."[24]

A critique no less devastating was published by the respected historian of astronomy Agnes M. Clerke in the *Edinburgh Review* for October 1896. Referring to the idea to which Campbell had also objected—that the canals carried surplus water across the equator far into the opposite hemisphere for purposes of irrigation there—Clerke maintained incisively that Lowell's theory, far from having been proved to a mathematical certainty, was on the contrary absurd:

> We can hardly imagine so shrewd a people as the irrigators of Thule and Hellas wasting labour, and the life-giving fluid, after so unprofitable a fashion. There is every reason to believe that the Martian snow-caps are quite flimsy structures. Their material might be called snow *soufflé*, since, owing to the small power of gravity on Mars, snow is almost three times lighter there than here. Consequently, its own weight can have very little effect in rendering it compact. Nor, indeed, is there time for much settling down. The calotte does not form until several months after the winter solstice, and it begins to melt, as a rule, shortly after the vernal equinox. . . . The snow lies on the ground, at the outside, a couple of months. At times it melts while still fresh fallen. . . .
>
> No attempt has yet been made to estimate the quantity of water derivable from the melting of one of these formations; yet the experiment is worth trying as a help towards defining ideas. . . . [We find that] only one-seventh of a foot of water, . . . could possibly be made available for their fertilisation, supposing them to get the entire advantage of the spring freshet. Upon a stint of less than two inches of water these fertile lands are expected to flourish and bear abundant crops; and since they completely enclose the polar area they are necessarily served first. The great emissaries for carrying off the surpluses of their aqueous riches would then appear to be superfluous constructions, nor is it likely that the share in those riches due to the canals and oases, intricately dividing up the wide, dry, continental plains, can ever be realised. . . . Further objections might be taken to Mr. Lowell's irrigation

scheme, but enough has been said to show that it is hopelessly unworkable.[25]

Of course, Lowell had allies as well, including—in a qualified way—the most illustrious Martian observer of them all, Schiaparelli. Even before Lowell's *Mars* appeared in print, the Italian astronomer had written probably his most provocative Martian paper, "La Vita sul Pianeta Marte" (The Life of the Planet Mars), which appeared on June 1, 1895.[26] In it he presents a closely reasoned interpretation of the Martian surface features, starting with the old idea of the dark areas as seas. He allows himself, moreover, the for him unusual liberty of speculating on the conditions of life on the planet. Flammarion, no conservative himself, was fairly shocked to find such a paper coming from such a quarter.

Schiaparelli begins by reflecting on the fact that on Earth the melting of the arctic and antarctic snow is of little real importance, because the polar glaciers are surrounded on all sides by an ocean extending from one pole to the other. Thus a rise in the water level due to the melting of part of the snow at one pole is reciprocated by the freezing of water in the other. On Mars, however, the distribution of seas is asymmetrical; though there is a great sea surrounding the south pole, says Schiaparelli, the north pole lies on a continent, and the only sea is that which is temporarily formed with the melting of the north polar ice cap. Moreover, the seas of each hemisphere are separated from those of the other by land masses, with no communication except by the canals. On Mars, therefore, the melting of either ice cap ought to produce changes in the hydrographic system of the planet of far greater consequence than on Earth, and indeed, Schiaparelli contends, this is just what is observed. As the snow melts, the waters rise and inundate all the shallow parts of the continents, and Schiaparelli concludes that on Mars "the periodic usurpation of the land by the sea resembles nothing so much as the great flux and reflux of our tides, which is no blessing for the country of Holland or the coastal areas of northwestern Germany, whose inhabitants must provide defenses which are assisted by dykes."[27]

Around the south pole, Schiaparelli continues, the dissolution of minerals into the great ocean located there, which occupies sometimes a full quarter of the Martian surface, must

make the water saline. Schiaparelli speculates that vast salt
beds may be found in the regions where the inundations take
place, though whether this is true or not, he adds, "in no case
would the water support cultivation of the land and the labor
of agriculture as is known among us." The temporary sea that
forms around the north polar cap as the snow there melts
becomes, then, of critical importance to the organic economy
of Mars. This temporary sea produces an inundation that "ex-
tends into numerous ramifications, and gives birth to vast
lakes" as it reaches toward the southern hemisphere and the
great ocean that constitutes the "natural basin of the Martian
waters." Since snow is the product of atmospheric distillation,
the water making up the northern sea is, unlike that of the
southern, fresh, and Schiaparelli concludes that "if there is any
organic life on the planet, its preservation depends above all
on this water."[28]

Schiaparelli pauses here to take stock of the differences between Mars as he has here portrayed it and the Earth:

> We are very privileged on Earth. The rains fall freely, and
> freely also the snow condenses on the summits of the
> mountains. The streams and rivers carry water to us
> effortlessly. The poor Martians have much harder condi-
> tions to endure. Rarely are there clouds, never rain. There
> are no fountains, no running water. The only resource is
> this great northern inundation; on it depends the whole
> question of existence. The ultimate prize is to capture this
> water before it is lost to the southern sea. . . . The life of
> the citizen depends upon it.[29]

Schiaparelli admits that with this he has entered the "realm
of romance" (though insisting that what he has been saying is
in reality no more bold or offensive than what others had
already published "under the sacred name of science"). With
this as an excuse, as it were, he turns his attention to the
"mysterious geminations and the extraordinary regularity with
which they take place." He points out that the idea that these
are best explained as owing to the activity of intelligent beings
"ought not to be regarded as an absurdity."[30] On the contrary,

> one cannot [otherwise] comprehend how in the same val-
> ley the moisture and vegetation sometimes make a single

line, in other cases two parallel lines of unequal breadth
and separated by unequal intervals, between which remains a sterile space deprived of water. Here, the inter-
vention of intelligent thought seems well indicated.[31]

Indeed, he himself, the "conservative" Schiaparelli, works out
an ingenious system of locks and dykes that might both regu-
late the water flow on Mars for the convenience of the inhabitants and account for the observations made from Earth.
(With his training in hydraulic engineering, it must have given
him some satisfaction to have the opportunity, if only on paper, to work out such a system for a whole planet.) Finally,
he concludes with some thoughts on the probable sociology
of Mars:

> The institution of a collective socialism would appear very
> likely to have resulted from one community of interests
> and one universal solidarity among the citizens, a true
> phalanx which could be considered a socialist paradise.
> One may well imagine a grand Federation of Humanity
> in which each valley constitutes an independent state. The
> interests of each individual and those of all are not to be
> separated. The mathematical sciences, meteorology, hy-
> drography, and the art of construction are without doubt
> brought to the highest degree of perfection. International
> disputes and wars are unknown; the efforts of all intellec-
> tuals, which among the insane inhabitants of their neigh-
> bor planet are wasted, are there directed against the
> common enemy: the difficulties which begrudging nature
> opposes to them every step of the way.[32]

There are certainly differences between this Schiaparellian
view of Mars and the one that Lowell was working out at
about the same time. Lowell, of course, had done away with
the identification of the dark areas with standing bodies of
water altogether, though the temporary seas forming as the
polar caps melted were no less important to his scheme.
Whereas Schiaparelli's Martians had confined themselves to
modifying natural features of the planet—valleys caused by
some geological process—Lowell's more superhuman race
were supposed to have designed and dug the whole system
from scratch. There was no great problem with this. In his

poem Lowell had written of the "strength phenomenal" of the
Martians. They were "more godlike than any Grecian god /
Who ever on Olympus trod." Yet their mental prowess was in
every way a match for their physical strength: their "mental
faculties transcend ours," he wrote, "as our own do the poor
brute's powers."[33]

While Schiaparelli regarded the Martians as probably having formed a socialist confederation, Lowell saw them as ruled
by an oligarchy of the elite. Yet in either case they were peace-
loving folk, quite unlike the initiators of interplanetary war-
fare about which H. G. Wells was soon to be writing in *The
War of the Worlds*. All in all, the similarities were far more
striking than the differences, leading E. M. Antoniadi to conclude—mistakenly, as it turns out—that because Schiaparelli
had first enunciated "the idea of the artificial origin of the
canals, conceiving the larger of them to be composed of six
different watercourses, whose dykes would be opened now
and then by the Martian minister of agriculture," Lowell and
others who proposed such theories were simply imitators.[34]
But in any event, Schiaparelli's paper seems to have been regarded, at least by the author himself, as something of a jest.
That is the implication of the comment he appended to the
end of the essay: "I leave now to any lecturer who cares to do
so to continue these considerations; as for me, I am descended
from a hippogriff." On the copy he sent Flammarion, he
scrawled at the top of the first page: "Semel in anno licet
insanire" (It is lawful to say insane things twice a year).[35] But
as before, when Schiaparelli had refused to "combat a supposition which includes nothing impossible," others were
more than willing to carry on where he had left off—and here,
at least, he seemed explicitly to be inviting it.

Not so surprisingly, perhaps, he later gave Lowell's theories
a sympathetic hearing, writing in 1897 that the system of
canals "presents an indescribable simplicity and symmetry
which cannot possibly be the work of chance," and writing to
Lowell himself: "Your theory of vegetation becomes more and
more probable."[36] Such comments lead one to believe that his
paper concerning "life on the planet Mars" had not been
wholly in the hippogriff mode. Yet in 1899, in a review of
Lowell's observations, he was striking the old note of agnosticism again, writing that the nature of the canals was still

"entirely obscure, despite the theories, oftentimes pretty and very ingenious, which they have occasioned."[37] It would appear from his last recorded utterances on the planet, in 1910, that whatever one wishes to make of his brief public flirtation with the idea of intelligent life on the planet, he died thus—an agnostic.[38]

THOSE WHOSE IDEAS we have been considering thus far in this chapter, whether accepting the idea of intelligent life on Mars or criticizing it, were basically in agreement on one major point: the objective existence of the canals as genuine features on the surface of the planet. Even Campbell, severe as he was toward Lowell, had highly praised Schiaparelli, and indeed had gone so far as to suggest, somewhat unfairly, that one of the real problems with Lowell's book had been that it was merely a rehashing of old ideas, in the main those of the great Italian astronomer himself.

Lowell's mathematics professor at Harvard, Benjamin Peirce, had begun his 1870 textbook *Linear Associative Algebra* with the sentence: "Mathematics is the science which draws necessary conclusions." Lowell may have believed that he had established the existence of life on Mars to a mathematical certainty. "The lines form a system," he wrote. "Instead of running anywhither, they join certain points to certain others, making thus, not a simple network, but one whose meshes connect centres directly with one another. . . . The intrinsic improbability of such a state of things arising from purely natural causes becomes evident on a moment's considera-tion."[39] Again, as Schiaparelli had written, "the system . . . cannot possibly be the work of chance."[40] Yet no argument, no matter how airtight its logic, is any more certain than the premises on which it is based—and where the canals were concerned, the observational basis was indeed anything but impregnable.

The year 1894, which saw Lowell enter the field, also saw the publication of an extraordinary paper from the hand of E. Walter Maunder, the English astronomer who had himself been one of the first to see the Schiaparellian canals.[41] It was this paper, "The Canals of Mars," that E. M. Antoniadi later had in mind when he spoke of Maunder as "the man who, by a masterly interpretation of facts, saw clearly into the 'canal' deadlock at a time when everything was darkness to all."[42] Here at last was someone whose keen insight was a match for the keen eyesight of the Italian observer.

Maunder is a good example of the way in which fame and oblivion are inequably distributed among scientists. Though his insights about Mars were indeed masterly, and though he accomplished a great deal of other work of the very first rank—to name but one of his achievements, he founded in 1890 the British Astronomical Association—he is relatively unknown today. Certainly he is far less familiar than Flammarion and Lowell, who were on the wrong side of the canal issue. This was partly, no doubt, because he lacked their flair; there was little of the realm of romance in his writings, unless one so considers his references to traditional Christian revelation. Partly this was also because he assiduously avoided the kind of personal attacks that characterize so much of the controversialist writing of the day. He was always civilized and was said to be "a man who never willfully said an unkind word."[43] Admirable as such qualities may be, they do not make for interesting press.[44]

The youngest son of a Wesleyan minister, Maunder was born in London in 1851. As a boy he showed something of his later abilities in a project he set for himself while recovering from a serious illness. He made a map of the district where he was living by pacing the length of the streets and measuring the angles by eye. After attending King's College, London, he worked for a time in a London bank, but his real chance came in 1873, when the position of photographic and spectroscopic assistant at the Royal Observatory at Greenwich became vacant. It was to be filled on the basis of performance on a civil service examination, and Maunder scored high enough to get the position. He was assigned to photographing sunspots and measuring their areas and positions on the solar disk, and in the course of this work he became impressed with the fact that "the smallest portion of the Sun's surface visible by us as a separate entity, even as a mathematical point, is yet really a wide extended area."[45] Thus he concluded that it was only reasonable to avoid taking too literally what appeared to be there—finding, for example, that the visibility or invisibility of spots did not provide sufficient grounds for estimating their areas, but to the contrary, that "it is often possible to see a straggling group of unimportant spots, when a single spot of considerably greater total area would be invisible."[46]

An observation somewhat similar to this had been made, incidentally, by Beer and Mädler long before in the course of their observations of the Moon. Calling attention to some "streaks of darkish material" they had noticed between the great craters Eratosthenes and Copernicus, they found that "when examined closely under favorable conditions, one detects here an innumerable host of the tiniest craters, whose shadows—individually indiscernible—blend together to form these fine streaks."[47]

Maunder pointed out that just as the smallest portion of the Sun's surface visible to us as a point was really a widely extended area, "the same truth applies in its degree to the planet Mars." He wrote: "We have no right to assume, and yet we do habitually assume, that our telescopes reveal to us the ultimate structure of the surface of the planet."[48] The tendency of the eye to link small disconnected spots might, moreover, even throw some light on the Martian canals:

> A narrow dark line *can* be seen when its breadth is far less than the diameter of the smallest visible dot. Further, a line of detached dots will produce the impression of a continuous line, if the dots be too small or close together for separate vision. There are some intimations that this may be the next phase of the "canal" question, Mr. Gale, of Paddington, New South Wales, having broken up one "canal" into a chain of "lakes" on a night of superb definition, Mars being near the zenith, and Prof. W. H. Pickering, at Arequipa, having under equally favourable conditions detected a vast number of small "lakes" in the general structure of the "canal system."[49]

Yet Maunder did not at this point decide between these alternatives—the canals as true lines or as roughly aligned dots drawn together by the eye. Indeed, he was actually less concerned with accounting for the canals than with accounting for the rapid changes in the boundaries of the larger markings, which it had been customary to account for—as Schiaparelli, for instance, had done—in terms of inundations of continental areas by seas:

> It seems a violent hypothesis to call in inundations extending over many thousands of square miles to account for merely temporary changes, for sooner or later the old districts take on the old configurations. . . . Indeed, it may happen that whilst several independent observers have recorded a change, others equally skilled have seen the planet as before.
>
> But if what we see is not the ultimate structure of the planet's surface, if, especially in the half-tone regions . . . we have an intermingling of minute areas of dark and light—be they water and land, forest and bare rock, prairie and sandy desert, or what you will—it is easy to see how enormous changes may apparently occur in a very little time. . . . When the difference in tone in two contiguous markings is but small, but a little defect in the transparency and steadiness of our own atmosphere will be sufficient to render them indistinguishable.[50]

One might well propose as a loose but just paraphrase of Maunder's main point St. Paul's famous passage in 1 Corinthians: "Now we see through a glass darkly." Maunder would probably have appreciated the comparison of his argument with Pauline doctrine, for he was a devout Christian who, following his retirement from the Greenwich Observatory in 1913, would become for a time secretary of the Victoria Institute, which was devoted to reconciling religion and science. He at one time wrote a book on the astronomy of the Bible that won the praise of a leading rabbi, of the pope, and of the archbishop of Canterbury, and his other astronomy books—he was a prolific writer—contain occasional passages such as this concerning the Galaxy: "It is nature at her vastest that we approach: we look up to her in her most exalted form. We see enrolled before us the volume which the finger of God has written; we stand in the dwelling-place of the Most High."[51]

Though some men of religious views—Father Secchi, for instance—had found no difficulty in reconciling Christianity with the idea of multitudinous inhabited worlds, others, such as William Whewell in his book *Of the Plurality of Worlds*

(1854) and Alfred Russel Wallace in *Man's Place in the Universe* (1903), had returned to the view of man's special place in God's creation. Wallace had argued that "the nearly central position of our Sun in the great star system" (when he was writing, this was still a tenable interpretation of the facts) was a reflection of man's divinely appointed centrality.[52] One suspects that Maunder may have been more in the camp of Wallace and Whewell than in that of Secchi. Indeed, his own 1913 book *Are the Planets Inhabited?* discusses the question in strictly scientific terms based on a consideration of the physical conditions of the planetary surfaces, but it comes to the same negative conclusion concerning the existence of planetary peers as had Whewell and Wallace. It may be, then, that Maunder had personal reasons for wishing to explain away the testimony of the system of canals vis-à-vis the existence of Martian life. If so, it must certainly be counted a small irony that the same telescope Lowell used in the hopes of proving the existence of life on Mars was later to be used by V. M. Slipher to discover the large red shifts of the spectral lines of what were then called white nebulae, which served as the basis of the discovery of the expanding universe and of a cosmology in which humankind's claim to a central position became not only untenable but absurd.

Initially, Maunder's article on the canals of Mars attracted little attention. He was simply a conservative grumbler, a maunderer, against an idea that bid fair to sweep all before it. This was the idea of intelligent life on another world. Lowell had seized on it and had given it what looked like definitive form. Thanks to his skillful advocacy, it appeared to be fast on the way to being accorded, in William Graves Hoyt's phrase, "conditional credulity."[53] Some of Lowell's observations, it is true, were soon to be called into serious question, but this involved Venus, not Mars.

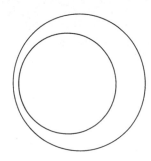

# CHAPTER 14

# A Stately
# Pleasure Dome

*In Xanadu did Kubla Khan*
*A stately pleasure dome decree.*
                    —Coleridge, "Kubla Khan"

THOUGH LOWELL was later to write that from Flagstaff "details invisible at the average observatory were presented at times with copper-plate distinctness, and, what is as vital, the markings were seen hour by hour, day by day, month by month,"[1] in fact conditions there during the winter of 1894—95 turned so bad that he actually considered abandoning the site for good. In view of the fact that Mars's next opposition would take place in midwinter, on December 1, 1896, it seemed hopeless to try to observe it from Flagstaff, where during the winter of 1894—95 Douglass found "not a single perfect night . . . and scarcely one or two which could be called good."[2]

Considering that Lowell was presently to go on record as saying that "the question of atmosphere" is the "*crux observationis*"[3] and that he would repeatedly excoriate other observatories for their poor atmospheric conditions, it is a bit surprising to find that the much-touted atmosphere at Flagstaff was conceded, even by Lowell himself, to be "not perfect; at times I am tempted to call it not even good,"[4] as he admitted privately in 1903. Though "details . . . were presented at times with copper-plate distinctness," the times involved were, in fact, mere fractions of seconds, as he makes explicit in his book *Mars*: "The steadiest air we can find is in a state of constant fluctuation. In consequence, revelations of detail come only to those who patiently watch for the few good moments among the many poor."[5] Yet though not wholly satisfied with conditions at Flagstaff, he found that it was not easy to improve on this substandard standard. He found conditions in the Sahara, which he visited early in 1896, to be mediocre,[6] and though he moved the observatory to Mexico for the 1896 opposition of Mars, afterward he moved it back to Flagstaff again. Other sites were tried in later years, but all in all Lowell seems to have become reconciled to the ups and downs of Flagstaff's seeing.

Lowell publicly boasted about the superiority of his conditions, whatever might have been his private reservations. Yet his constant need to employ diaphragms with his new 24-inch Clark refractor, used from 1896 on, makes one wonder. Generally he found it necessary to "gag" the instrument to 12 to 18 inches because the atmosphere did not permit use of the full aperture. Whereas Young at Princeton had found with the 23-inch refractor of the Halsted Observatory that the relation of maldefinition to aperture was "something like the square root," Lowell's experience tended rather to bear out Lord Rosse's cubic relation, established on the basis of experience with very large reflectors in the atmospheric conditions of England, which were not famous for steadiness. Moreover, he gave some idea of how infrequently he was able to use the full 24-inch aperture when he wrote in 1905 that

> practice in this art of assisting nature [with diaphragms] has shown it to be much oftener needful or beneficial than one would antecedentally suppose. Incidentally it enables an estimate to be formed of the relative proportion of times at which a glass of a given size can profitably be employed for planetary work. This proves to be inversely as something between the square and cube of its diameter. So that the disadvantages connected with atmospheric waves far exceed the advantages to be derived from the light waves, especially in the matter of continuous work.[7]

E. M. Antoniadi, who used a still larger aperture for planetary observation at Meudon a few years later, came to the conclusion that Lowell "exaggerated the importance of the role of atmosphere with regard to powerful instruments." He noted that whereas Lowell claimed "absolute superiority" of observing conditions at Flagstaff, even in northern France

**Figure 14.1**
*Lowell at the telescope, October 17,
1914.*

conditions were sometimes perfect and that in 1909 in particular it was possible to use the 32³/₄-inch refractor without a diaphragm; it then showed Mars as if twice as near as seen from Flagstaff with the 24-inch diaphragmed to 16 inches.[8]

The Lowell 24-inch was, in the opinion of its maker, Alvan G. Clark, optically as perfect an instrument as the firm had ever made. Intended as the observatory's permanent telescope—replacing the borrowed 18-inch Brashear that had by now been returned to Pennsylvania—it had been built short enough to allow it to be fitted into the dome built for the smaller instrument in 1894. Even so, it was a tight fit, the lens having to be hoisted up and passed through the dome's shutter. Installed at Flagstaff in July 1896, the telescope was trained on Mercury and Venus by Lowell and his assistants in August, and their surfaces at once revealed various markings, of which those on Venus in particular came to embroil Lowell in a heated and protracted controversy.[9] At about the same time, T. J. J. See, then considered a skillful observer of double stars, reported the recovery of the faint companion of Sirius, lost in the glare of the brighter star since 1890. It eluded other observers, apparently testifying to the excellence of Lowell's glass and atmosphere. However, in November J. M. Schaeberle and R. G. Aitken recovered it with the Lick 36-inch refractor in a different position from that given at Flagstaff.[10]

Making the Lowell refractor short enough to fit into a dome built for a smaller telescope may itself have compromised its effectiveness. The telescope's focal ratio is, at f/16, somewhat shorter than that of most telescopes distinguished for planetary work. The problem with a shorter focal ratio for a large refractor is that the deleterious effects of chromatic aberration tend to be increased, for even "achromatic" lenses are only relatively free of false color. Antoniadi found the Meudon telescope (f/19.3) to be "free of chromatic aberration" when he observed Mars in 1909, but Clyde Tombaugh alleges that, based on long experience with the Lowell refractor, chromatic aberration is such a serious problem with the latter instrument that "I found no gain of planetary detail when using more than 20 inches diameter aperture. Indeed beyond 20 inches I saw less detail."[11] The remedy, in any case, was the same as that proposed for atmospheric disturbance: diaphragming the lens. (Significantly, when diaphragmed to 20 inches or so, the Lowell 24-inch acquires an effective focal ratio of about f/19.) It would be quite unfair to say, as Antoniadi later did, that when Lowell's powerful telescope "showed too much on [Mars], with irregular markings, he found it necessary to draw nearer the conditions of [Schiaparelli's] observations in order to see the rectilinear canals, and therefore diaphragmed the objective."[12] In fact, Lowell had sound reasons for diaphragming *his* objective. Yet it certainly must be admitted that the net effect was that, despite having now a larger telescope, Lowell was forced for one reason or another to use it as a smaller one, very nearly duplicating, in the customary conditions under which he observed with it, the conditions of observation that had shown the canals so agreeably during 1894. Nor had he any right to assume that results obtained with his particular instrument would be universally applicable to other instruments.

Lowell moved his telescope to Tacubaya, Mexico, near Mexico City, at the end of 1896. As far as the opposition of Mars that year was concerned, the move was a washout. Douglass, it is true, succeeded as in 1894 in making out canals in the dark areas, and Lowell—though only able to see a few of these himself—seized on their importance for his theory. As he wrote in the article "Markings in the Syrtis Major" concerning that most conspicuous of Martian dark areas (and the one Huygens had used to first measure the planet's rotation period in 1659),

It was very early surmised to be an ocean, and from its shape was called the Hour-glass Sea, a name of doubly happy significance since it had served as Martian hour-glass to observers on Earth. A sea it continued to be considered by generations of astronomers. For over two hundred years its supposed aquatic character passed current practically unquestioned. Its color seemed in fact conclusive of its constitution. Under the best observational conditions it showed of a deep blue-green; just the hue an ocean might be expected to exhibit. . . .

Unfortunately for the ocean-loving the character of the Syrtis Major must pass with other charming myths into

the limbo of the past. For the great blue-green area is no
ocean, no sea, no anything connected with water, but
something very far removed from water: namely a vast
track of vegetation.[13]

In his vegetation theory Lowell made much of supposed
changes in the tints and intensities of the dark areas, but the
blue-green tints have now been proved to have a physiological
origin; a neutral-colored area surrounded by a field of bright
orange or red appears to have a blue-green tinge. Both the
dark and bright areas are actually reddish.

Though the results on Mars hardly justified the trouble of
moving the observatory to Mexico—and indeed Lowell himself did not arrive at Tacubaya until well past opposition—
Mercury and Venus showed once again the markings that had
first been recognized at Flagstaff in July. Observations were
made by Lowell and all of his assistants, who included, at the
time, besides Douglass and See, Wilbur A. Cogshall, Daniel
A. Drew, and Lowell's secretary, Wrexie Louise Leonard. W.
H. Pickering, who had been with Lowell in 1894, had meanwhile returned to Harvard following expiration of his year-
long leave of absence from that institution.

LOWELL COVERED Mercury with a Byzantine network of
lines (see Fig. 7.3), and indeed he took pains to stress that
these were *not* artificial-looking like those on Mars but in all
likelihood were cracks in a surface baked by the intense heat
of the Sun. The alleged differences between the tessellations of
his map of that planet and those then shown on maps of Mars
were, however, perhaps obvious to few besides himself.

Yet it was his Venus work that really strained credulity to
the limit (Fig. 14.2). He found on its surface a set of markings

surprisingly distinct; in the matter of contrast, as accentu-
ated, in good seeing, as the [naked-eye] markings on the
Moon and owing to their character much easier to draw;
in the matter of contour, perfectly defined throughout,
their edges being well marked and their surfaces well dif-
ferentiated in tone from one another, some being much
darker than others. They are rather lines than spots. . . .
A large number of them, but by no means all, radiate like
spokes from a certain centre.[14]

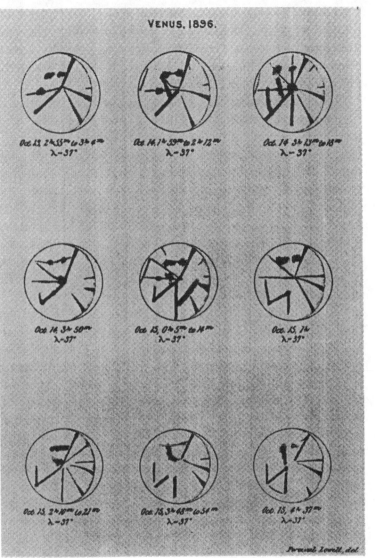

VENUS, 1896.

*Figure 14.2*
*Lowell's drawings of Venus, 1896.*

They were, to Lowell's eye, "so evident . . . that it has been possible to get position angles of them," and only when the seeing was very poor were they obliterated. Not only Lowell himself but all of his assistants made them out, "to say nothing of several outsiders," Lowell later alleged.[15]

The markings were "always seen in the same place" and appeared to confirm the synchronous rotation of Venus that had been announced a few years earlier by Schiaparelli (strange, though, that the center of Lowell's system of spokes should directly face the Earth rather than the Sun). Moreover, the markings were apparently on the surface itself, despite the obvious objection that nearly every astronomer of the day was convinced that Venus was surrounded with highly reflective clouds. Lowell later admitted that "the high albedo of the disk is . . . a difficulty which must be accounted for," but he added, "it seems possible to do so by the idea of a bright atmospheric veil as well as by a cloud canopy."[16]

Various diffuse patches had been reported on the planet since the invention of the telescope, as we have seen. But there was always the question of whether these were genuine. Schiaparelli, summarizing the work of his predecessors, had deemed most of it worthless, and the skilled observer E. M. Antoniadi concluded in 1897 from his own studies that "the observation of the diffuse shadings of Venus is of enormous difficulty, and I cannot consider myself authorised to say that I have seen well-defined spots on this planet." Moreover, though "certain shadows . . . were sometimes sufficiently marked to allow of their being represented on paper," he had little confidence in their reality, tending to regard them as contrast effects.[17] Indeed, because of its dazzling sheen, no planet is more liable to produce optical deceptions. Its intense lustre highlights every imperfection of the telescope, and of the eye that uses it. On this last point, some comments written by A. E. Douglass in his article "Atmosphere, Telescope, and Observer," dated from Mexico City on April 2, 1897, seem apropos:

> Perhaps the most harmful imperfection in the eye is the lack of homogeneity within the more dense transmitting media, either the lens or membranes, probably the former. Under proper conditions the lens (presumably) displays irregular circles and *radial lines,* the whole resembling a

*spider-web structure. . . .* Under actual tests this structure is so very prominent that we wonder how the eye is able to give such good definition as it does. No optician could ever sell a lens so badly made, except for the coarsest usage; in proportion to its size it has the imperfections one finds in the lens of a bull's-eye lantern. (Italics mine.)[18]

One wonders about the possible relation to Lowell's "spokes" of the "radial lines" and "spider-web structure" that Douglass found so prominent under the proper conditions, though unfortunately Douglass fails to address this question.

There was, in any event, certainly a long tradition of deserved skepticism with respect to this planet, and it could hardly be expected that Lowell's markings would be allowed to pass by unscathed. In fact, the criticism was fairly severe. Antoniadi wrote sarcastically: "I refrain from discussing the aphroditographic work of most of our contemporaries, who forgetting that Venus is decently clad in a dense atmospheric mantle, cover what they call the 'surface' of the unfortunate planet with the fashionable canal network." Referring to the tendency of Lowell's markings to remain always directly facing the Earth, he added that "such discussion would lead us to some curious conclusions in which we would not be long in finding that Venus does not constantly present the same face to the sun but to the earth."[19] E. S. Holden, at Lick, ascribed Lowell's class of markings to "illusions of some sort," and proposed that perhaps they were due to strains in the object glass produced by setting the adjusting screws too tightly.[20] Holden noted that "all observers, except those at Flagstaff, see markings of one class; while those drawn by Mr. Lowell are of a totally different nature." This was not quite true. There have been a few confirmations of Lowell's work then and since. A Mr. Foulkes at Malta reported that he had "the greatest faith" in Lowell and that he was quite willing to accept the "canal-like markings," having often seen himself "a manner of very faint streaks, which I could not locate, but they were there." He added, with what sounds like a note of sarcasm, though perhaps it was not intended to be: "I never ventured to try to put them on any of my drawings, because I make it my rule to insert only what I really see."[21]

A few amateurs, using various little telescopes, have claimed to see Lowell's markings over the years, and there is a resemblance between Lowell's chart and that produced by a professional astronomer, Audouin Dollfus, a planetary observer of undoubted skill.[22] Of those on Lowell's staff who saw the markings, only Miss Leonard actually approached his boldness in depicting them, though See and Douglass both took up the cudgel, at times, in their defense. Douglass wrote a spirited reply to the skeptics in which he revealed some interesting facts about the circumstances under which the observations had been obtained. The markings were not, it turned out, seen with anything even approaching the full aperture of the 24-inch, the planet being then "too bright and badly 'shattered' by air currents." Instead, the detail had been "perfectly evident" only when the aperture used was from 1.6 to 3 inches! Douglass concluded that "when a man has had a large experience under particularly favorable circumstances, like Mr. Lowell, his report is not lightly to be set aside," and he illustrated this by his own experiences in the detection of a new class of markings on Jupiter's satellites:

> The first reason why other observers have not seen these markings is bad atmosphere. When I began observing the third satellite of Jupiter, for days, even weeks, I drew nothing but hazy indefinite markings or belts, such things as M. Antoniadi describes as appearing to him on Venus. But one night after making several drawings of that character the seeing suddenly became superb, the curtain rose as it were, and I saw sharp distinct black lines about which uncertainty was impossible. The very same thing happened on the fourth satellite four days later. I had been drawing the same indefinable shadings, when one night the seeing improved, the curtain again rose, and I perceived sharp definite lines. And once thoroughly understanding the character of the object sought, I could see them and profitably study them under conditions of seeing formerly prohibitory.[23]

Needless to say, Douglass's lines on the satellites of Jupiter were to prove no more immune to criticism than the observations he had attempted to defend, and indeed the tendency of the Flagstaff observers to cover every planetary surface in sight

with lines no doubt unwittingly worked to undermine confidence in the canals of Mars. Much as Lowell protested that the
Martian canals were in a class by themselves, it was evident
that they were not quite as unique as had been supposed, and
that the most likely explanation for the family resemblance was
that they were *all* illusory.

Planetary astronomy was becoming a bit of a mess. E. E.
Barnard wrote with respect to Douglass's observations of the
satellites of Jupiter:

> As I failed to see these markings but saw others not
> shown by Mr. Douglass, it becomes a question of tele-
> scope and atmospheric conditions as to why he should see
> one thing and I should see another so wholly different in
> appearance.
>
> I believe it to be a fact that no telescope existing—
> unless of course it is the Yerkes 40-inch when it is in
> working order—is so capable of showing the surface fea-
> tures of these satellites as the Lick 36-inch. I also believe
> there are brief intervals of good seeing at the Lick Obser-
> vatory which are not excelled if indeed they are equalled
> at any other observatory. . . . Though I have observed
> and drawn markings on these satellites, I have not seen
> any narrow straight lines or similar markings on any of
> them. The markings I have seen have always appeared to
> be large and more or less diffused—except in the case of
> certain white polar spots.[24]

Barnard went on to criticize, moreover, the prevalent drawing style of observers of the day, which had begun with Schiaparelli's diagrammatic representations of Mars and, in the
hands of Lowell and his assistants, was being extended wholesale to the rest of the solar system:

> It has always appeared to me that when representing any
> planetary detail it should be drawn as nearly accurate, as
> to appearance, as possible. If the marking is vague and
> uncertain it should be made so on the drawing and no
> definite outlines should be given where none exist. Some
> observers seem to be in the habit of giving a definite
> boundary to markings whether such are really seen or
> not. This is very misleading. . . .

This diagram method of drawing is specially noticeable in a good many drawings of Mars, and it is perhaps through the study of these and not by any inspection of the planet itself that so many queer and unnatural ideas have been propagated concerning the physical appearance of Mars. Many of the regions shown on drawings and maps of Mars as definitely bounded, are in reality very diffused and uncertain in their outlines.[25]

Such was the pronouncement of this skilled observer using what was then the largest instrument in existence. Douglass seems to have taken the censure to heart, for not much later he became skeptical not only of the linear details on Jupiter's satellites but also of those on the other planets as well. He began to wonder about the extent to which imitation of one's own or others' drawings played a role in what was recorded on planetary surfaces.

DOUGLASS WAS, at this time, acting director of the Lowell Observatory, as Lowell himself had been forced to give up astronomical work for the time being owing to "neurasthenia." While still in Mexico he had begun to suffer from nervousness and fatigue, and though after his return to Boston he had set out by train for Flagstaff (to which the observatory had by now been returned), he got only as far as Chicago and then turned back. His brother, according to the prevalent notions of the day, attributed the onset of illness to the strain of overwork:

> Although he could stand observing day and night without sufficient sleep while stimulated by the quest, the long strain proved too much, and he came back to Boston nervously shattered. Such condition is not infrequent with scholars who work at high speed, and although the diagnosis is simple the treatment is uncertain. The physician put him in bed for a month in his father's house in Brookline.[26]

By midsummer Lowell had apparently improved enough to write Douglass, "My quantity improves and they tell me my quality will come after it."[27] In fact, it would be four years before he felt well enough to return to Flagstaff.

The term *neurasthenia* had been coined by New York neu-
rologist George Miller Beard in the 1860s to describe a rather
amorphous syndrome typically consisting of anxiety, fatigue,
insomnia, and somatic complaints, and thought to be a form
of nervous exhaustion brought on by the stresses of modern
industrialized life.[28] As a diagnostic category it was too broad
to remain intact indefinitely, and it has now been all but abandoned; probably most of those who were so diagnosed were
suffering from what we would today describe as anxiety disor-
ders or mild depression.[29] But for a while the diagnosis was all
the rage. Beard alluded to its "contagious quality." Four-fifths
of his neurasthenics came from what he described as the
higher order of American society; a full quarter were physi-
cians. It became, writes George Frederick Drinka in *The Birth
of Neurosis,* "the malady of the great, the sensitive, the over-
worked, the 'brainworkers'—all those at the ethereal tip of the
social evolutionary ladder. . . . It is a wonder that all West-
erners did not take to their beds and refuse to get up, blaming
everything on neurasthenia."[30] Overwork was generally considered responsible for the illness, and in fact Lowell wrote
that his illness represented "a complete breakdown of the machine."[31] William James, another neurasthenic, whom Lowell
met in April 1900 on the French Riviera, where both were
recovering, had written in 1873:

> Twenty-five years ago, clerks and young employees hardly
> ever expected a holiday, except as a matter of particular
> favor. . . . That was a time when we lived under the dis-
> pensation of the favorite American proverb—no half truth
> even, but an invention of Sabbathless and unvacationed
> Satan—"Better to wear out than rust out." But of those
> who repeated it with most faith, how many have since
> had enforced leisure to repent their short-sighted-
> ness. . . . Who that has traveled in Europe is not familiar
> with the type of the broken-down American businessman,
> sent abroad to recruit his collapsed nervous system? With
> his haggard, hungry mien, unfitted by life-long habit for
> taking any pleasure in passive contemplation, . . . these
> Americans have been brought up to measure a man solely
> by what he effects, hardly at all by what he is.[32]

We of the post-Freudian era cannot help noting that there were, in the breakdowns, obvious secondary gains. The breakdown and its treatment (rest and electrotherapy were the most common) provided a temporary reprieve from the patient's huge self-demand; if it was a regression of sorts, a retreat to childlike dependency, at least it was a socially acceptable one.

Alice James, who like her brother William was diagnosed as suffering from neurasthenia, recognized in herself repressed feelings of hatred toward her father that at times became parricidal. "I saw so distinctly," she wrote in her diary, "that it was a fight simply between my body and my will. . . . Owing to some physical weakness, excess of nervous susceptibility, the moral power *pauses,* as it were for a moment, and refuses to maintain muscular sanity, worn out with the strain of its constabulatory functions."[33] The implication is that the "nervous exhaustion" was not, after all, owing to overwork but to the strain of the will's "constabulatory functions" in holding back feelings that were unacceptable to the ego. While parricidal is a strong word, it is easy to accept that the neurasthenic's intermittent work paralyses may have had something to do with a deep sense of the inadequacy of one's general equipment in the face of the demands of an overdeveloped superego (in Lowell's case founded, presumably, on his relationship with his martinet father). Lowell would seem to offer a promising case for retrospective psychoanalysis, and though his unconscious conflicts have yet to be thoroughly appraised, one can agree with at least this much of analyst C. K. Hofling's preliminary attempt: that they probably had a great deal to do with his relationship with his father.[34] On the other hand, his periods of intense activity alternating with apparent depressions are extremely suggestive of manic-depressive illness, which is not infrequently encountered among highly creative individuals.

Lowell's relationship with his mother seems to have been tenderer; they were certainly very close, and even when the observatory was going up at Flagstaff in 1894 he wrote her nearly everyday. Psychoanalysts have tended to see the child's loss of unity with the mother—whether with the discontinuation of suckling at the breast or with the diversion of attention to a younger sibling—as producing a dim but universal nostalgia for a lost paradise. It is at least tempting to see in the vision of the lost garden that informed Lowell's Martian theories some reflection of a lost paradise he had personally known, just as the "pleasure dome," incarnated in his observatory, was perhaps symbolic of the mother's breast. All this is, of course, sheer speculation, and it does not promise to lead very far. But there is often something to be gleaned from a person's earliest memory. In Lowell's case, this was of Donati's Comet of 1858, and Lowell recalled years later: "I can see yet a small boy half way up a turning staircase gazing with all his soul into the evening sky where the stranger stood."[35] Is it asking too much to think that the emotion he felt was not only toward this stranger in the heavens but also toward that other stranger that had entered his life at the time?

Nostalgia, whatever its origin, was in any case characteristic not only of Lowell but of the century as a whole. The importance of work—the need to be practical, to be "real"—is symbolized by the textile mills of Lowell's grandfather, or by schools such as the Lawrence Academy, which had been founded by that same grandfather for the express purpose of teaching science and practical technology (engineering, mining, and mechanical drawing). Yet beneath the progressive rhetoric of the Social Darwinist faith that industrial society was on the march toward better and better things, there was, perhaps not surprisingly, an unconscious desire to regress, a craving for something else that nineteenth-century man believed earlier ages had been able to take for granted but that in the present had been taken away, largely as a result of technical advance. Thus the Victorians already felt very keenly the modern ambivalence about technological progress, and adherence to its banner was mingled with distrust. Though faith in the old religious forms had waned, a desire to believe in something remained.

The lapse of old faiths becomes, in fact, one of the memorable themes of the century, expressed by Tennyson in his poem *In Memoriam* and by Matthew Arnold in "Dover Beach." The nostalgia for a time now lost is reflected in Henry Adams's *Mont-Saint-Michel and Chartres* and in "The Cathedral," a poem written by Percival Lowell's cousin James Russell Lowell, which speaks of

This nineteenth century with its knife and glass
That make thought physical, and thrust far off
The Heaven, so neighborly with man of old,
To voids sparse-sown with alienated stars.

. . . . . . . . . . . . . . . .

At best resolving some new nebula,
Or blurring some fixed-star of hope to mist.[36]

As in these last-mentioned works, the Middle Ages in particular were seen as an idealized time when "faith was at the full."
Percival Lowell was not immune to their attractions. In Japan
he enjoyed "sitting by the window and looking at the old
feudal remains below, the moat with its stagnant slime and the
red dragon flies skimming its surface, the old walls, the over-
grown ramparts. . . . All tended to carry my thoughts back to
the middle ages, or was it only to my own boyhood when the
name *middle ages* almost stood for fairy land?"[37] In Mars,
Lowell had discovered yet another world that was, if not in
ruins, well on the way to it. One wonders whether in the
fantasy he projected he was not again merely projecting the
bittersweet feelings of his own view of his forever irretrievable
boyhood.

Science, with its constant insistence on reality, on objectifi-
able fact, had spoiled for the nineteenth century the ability to
take refuge in the childish romances in which earlier ages (the
Middle Ages in particular) had found such comfort. The objective physical world that science had revealed seemed, moreover, increasingly alien to spiritual longings. The dead universe of matter in motion had seemed for some time to be a
stifling prison of the human spirit, as the body had been to the
Stoics. The puritanical Lowell, with his keenly developed con-
science and sense of guilt, evidently shared some of this dis-
gust with the body in the narrower sense. In the lines we have
quoted earlier from the poem "Mars,"

My far-off goal seems strangely near,
Luring imagination on,
Beckoning body to be gone
To ruddy-earthed, blue-oceaned Mars,

one wonders whether the double meaning can have been quite
unintentional; but even if unconscious, it intimates that his
sympathy lies with the ideal rather than with the physical.

The nineteenth-century reaction to the spiritual crisis took various forms. Some, like Cardinal Newman, sought the answer in a return to orthodox faith; others sought out unorthodox faiths. The interest in psychic phenomena and spiritualism that arose at the end of the century is characteristic, and some of the same minds that felt so keenly the attraction of the idea of extraterrestrial life felt, as we have seen, the attraction of spiritualism—Schiaparelli and Flammarion, for instance. While Lowell seems to have had little interest in orthodox religion, he too was intrigued for a time by psychic phenomena, and he seems to have been fascinated with the perplexing questions with which the spiritualists concerned themselves. He wrote in *The Soul of the Far East*:

> But however strong the conviction now of one's individuality, is there aught to assure him of its continuance beyond the confines of its present life? . . . Close upon the heels of the existing consciousness of self treads the shadow-like doubt of its hereafter. Will analogy help to answer the grewsome [*sic*] riddle of the Sphinx? Are the laws we have learned to be true for matter true also for mind? Matter we now know is indestructible; yet the form of it with which we once were so fondly familiar vanishes never to return. Is a like fate to be the lot of the soul? That mind should be capable of annihilation is as inconceivable as that matter should cease to be. Surely the spirit we feel existing round about us on every side now has been from ever, and will be for ever to come. But that portion of it which we each know as self, is it not like to a drop of rain seen in its falling through the air? Indistinguishable the particle was in the cloud whence it came; indistinguishable it will become again in the ocean whither it is bound. Its personality is but its passing phase from a vast impersonal on the one hand to an equally vast impersonal on the other.[38]

The spiritualists and extraterrestrialists had, moreover, a thoroughly nineteenth-century outlook in their concern with what might be called the verificatory aspects of their beliefs. Whereas in earlier ages it had been enough to create a visionary world that was psychologically satisfying, the upholders of these new faiths felt the need for their visions to be somehow *externally* validated with empirical methods—thus the profound interest in photographs of spirits, or later in photographs of the Martian canals. In the case of the spirit photographs, even Sir Arthur Conan Doyle was fooled. To us, they appear only too obviously faked, and we can account for the fact that astute observers were taken in only by their *willingness* to be taken in.

The new twist of the nineteenth-century faiths was the emphasis on the verificatory aspects. Freud defined illusion as "a belief derived from human wishes."[39] This did not mean, he emphasized, that it must necessarily be an error, or the same thing as an error, but only that "the underlying motive of the belief is some wish-fulfillment, and so we disregard its relation to reality, just as the illusion itself sets no store on verification."[40] While true of the faiths of the Middle Ages, perhaps, this was no longer true of the faiths of the nineteenth century, which set considerable store on just such verification. Yet with this qualification, they were no less illusions in Freud's sense: beliefs whose underlying motive was some wish fulfillment. Lowell wrote: "As we grow older we demand reality, but so this requisite be fulfilled the stranger the realization the better we are pleased."[41] Clearly this applied as much to mankind as a whole as to any individual; as *it* grew older, it too demanded reality, but the spiritual longings remained those of its infancy. Lowell had said of the impulse that lay behind the belief in Martians:

> We all have felt this impulse in our childhood as our ancestors did before us, when they conjured goblins and spirits from the vasty void, and if our energy continue we never cease to feel its force through life. We but exchange, as our years increase, the romance of fiction for the more thrilling romance of fact.[42]

There is a literary allusion here to Shakespeare's *1 Henry IV,* where the magician Owen Glendower boasts to Hotspur: "I can call spirits from the vasty deep." Lowell omits Hotspur's reply: "Why, so can I, or so can any man, / But will they come when you do call for them?"

Lowell's Martian theory had begun, after all, with a poem —a work of art rather than of science. As a young man uncertain of his direction in life, Lowell had intended to make

literature his field; one wonders whether this was another of
the choices about which he had felt "anxiety to whether his
father would approve."⁴³ He felt obliged to be practical, to be
"real." And so the poem became, in a sense, a disowned child;
its existence was never acknowledged, it was never brought
before the public in precisely this form. Lowell was trying
hard to be a man of science now, just as when, after his graduation from Harvard, he had set aside mathematics—which like
a work of art is a world unto itself with no need to make any
claim on reality—and attempted to become a man of business.
He gave the impression that his Martian theory had been
based on careful observation, but the observations were a mere
pretext. In fact, his theory was, as he said of mathematics,
simply "one vast imagination based on a few so-called ax-
ioms."⁴⁴ His assistant, Douglass, who during the four years of
Lowell's absence from the observatory gradually seems to have
freed himself sufficiently from his charismatic personality to
begin to see his work with some objectivity, wrote: "He de-
votes his energy to hunting up a few facts in support of some
speculation."⁴⁵ Others at Flagstaff, and the world generally,
were to remain fascinated not so much with Mars and what
Lowell was supposedly discovering there, as they thought, but
with Lowell himself. And no one was more fascinated with
Lowell than Lowell. Observation is best advanced through
making the record as impersonal as possible. Lowell, on the
other hand, put the stamp of his forceful personality on everything he did, not least of all on his observations. Just as the
poet vivifies all around him until the Moon takes on a countenance, Lowell—for poet he was—turned each planet into a
reflection of himself. Yet he had himself almost anticipated
that it might be so, writing from the Far East to a friend,
Frederic J. Stimson:

> Somebody wrote me the other day apropos of what I may
> or may not write, that facts not reflections were the thing.
> Facts not reflections indeed! Why, that is what most
> pleases mankind from the philosopher to the fair, one's
> own reflections on or from things. Are we to forego the
> splendor of the French salon which returns us beauty
> from a score of different points of view from its mirrors
> more brilliant than their golden settings. The fact gives us

but a flat image. It is our reflexions upon it that make it a solid truth. For every truth is many sided. It has many aspects. We know now what was long unknown, that true seeing is done with the mind from the comparatively meagre material supplied us by the eye.[46]

The critics of Lowell's Martian theories might well have made a telling point or two had they only known of this letter. One can only wonder whether he ever outgrew the childlike narcissism which it expresses. "Facts not reflections indeed!"—a reasonable attitude to take for a budding imaginative writer, perhaps, but less safely endorsed by an aspirant to the self-effacing methods of empirical science.

O F   T H O S E who observed Venus with Lowell in 1896–97, only his secretary, Wrexie Leonard, drew the markings with a boldness approaching his own, and indeed their drawings are sufficiently alike as to be almost interchangeable. The role of suggestion—the influence of one observer's seeing on another's—was one of the psychological questions that Douglass would soon be asking about planetary observation. But already in his attempted defense of Lowell's Venus work before he turned critical of it himself, Douglass had implicitly acknowledged the relevance of the question. "I had never closely studied Mr. Lowell's map—merely glanced at it casually," he wrote, "and though I knew it resembled the hub and spokes of a wheel, I did not know what position the centre held with respect to the phase . . . in order to give even a remote resemblance to Mr. Lowell's work."[47]

The mental susceptibility that, in William James's words, makes us "yield assent to outward suggestion, affirm what we strongly conceive, and act in accordance with what we are led to expect," is probably common to us all.[48] In closed settings such as convents it is known to have a tendency to produce, from time to time, episodes of mental contagion referred to as mass hysteria. In the hospital of the Salpêtrière in Paris, headed by Charcot—a man described by one who knew him as the "most authoritarian man I have ever known"[49]—it produced dramatic but, as we now know, fictitious convulsions in patients who were hypnotized, providing in the process striking evidence for Charcot's theories. Salpêtrière was also a closed system, and moreover,

> because of Charcot's paternalistic attitude and his despotic treatment of students, his staff never dared contradict him; they therefore showed him what they believed he wanted to see. After rehearsing the demonstrations, they showed the subject to Charcot, who was careless enough to discuss the case in the patients' presence. A peculiar atmosphere of mutual suggestion developed between Charcot, his collaborators, and his patients, which would certainly be worthy of an accurate sociological analysis.[50]

Lowell had witnessed what was in some ways an analogous phenomenon himself, in the case of the "fox disease" of Japan, "a species of acute mania supposed by the people to be a bewitchment by the fox. As the person possessed so regards it and others assist in keeping up the delusion by interpreting favorably to their own views, it is no wonder the superstition survives."[51]

To what extent was there such a "peculiar atmosphere of mutual suggestion" at Flagstaff during the years of Lowell's directorship? No doubt his assistants at times were given to accommodating themselves to Lowell's views, though perhaps less consciously than Charcot's assistants. When, as discussed below in chapter 15, Douglass raised serious questions about Lowell's work, he was dismissed as untrustworthy. He had spoken of Lowell's "strong personality, consisting chiefly of immensely strong convictions," which had, he said, led him (Douglass) to support ideas that he later regretted.[52] By that time the "spokes" of Venus and the lines on Jupiter's satellites were among the regretted ideas. Yet this was Douglass's view in hindsight. At the time, he had evidently been persuaded of the reality of such features. Lowell's coercion over his staff was no doubt great, but it was not as overt as Charcot's; it was the more subtle product of a charismatic personality that naturally invited the belief and allegience of others. Of course, this does not at all change the result: those who worked with him no doubt showed him more or less what they believed he wanted to see, but because they themselves came to want to see and believe in it.

Of all the members of Lowell's staff, there can be little question that Leonard was among the most affected by suggestion, if one can safely judge from her drawings. She was also the most captivated by Lowell's personality. It is hardly an exaggeration to say that she worshipped him, and she gives ample evidence of her continuing attachment to him even after his death in the collection of memorabilia to which she gave the title *Percival Lowell: An Afterglow,* which was published in 1921 in Boston, where she had moved after being dismissed by Lowell's widow shortly after Lowell's death. Perhaps Mrs. Lowell had been unable to tolerate the continued presence at the observatory of a woman who had at one time been Lowell's mistress. Indeed, one of the Lowell observatory astronomers told the late William Graves Hoyt that it was generally wondered on the staff why Lowell, at the age of fifty-three, decided to marry his Boston neighbor, Constance Savage Keith, rather than Leonard. Presumably considerations of social class entered in, but in any case Lowell continued to value his loyal secretary sufficiently to keep her on the hill despite what must have been in certain respects an awkward situation.

After Douglass was dismissed in 1901, Lowell gathered a new set of assistants around him. Competent, even brilliant, these included the Slipher brothers, V. M. and E. C., and C. O. Lampland. They were to prove unquestionably loyal to Lowell, both during his lifetime and afterward.

Lowell's observatory did not quite become a walled-off garden, however. There were visitors from the outside, though most of those who traipsed up the hill to the telescope for a look (and of whose "independent" confirmations Lowell made much) were only too willing to oblige him in seeing what he told them was there. These included his friends Lester Frank Ward, Edward S. Morse, and George Russell Agassiz (the man who had been with him when he first decided to make a career of astronomy). Robert Wheeler Willson, visiting the observatory in 1911, took turns with Lowell in looking at Mars through the new 40-inch reflector, and as Lowell wrote to Leonard, "He has seen many canals ill and one canal well. He feels he is getting on. He always believed but now is seeing for himself."[53]

*Figure 14.3*
*Dome of the 24-inch refractor of*
*the Lowell Observatory at*
*Flagstaff, Arizona.*

Another man sympathetic to Lowell who came to the observatory was the British amateur James H. Worthington, who observed both Mars and Venus with the 24-inch refractor in 1909. He wrote of his impressions of Venus at the time: "I observed the planet under favourable circumstances, and was able to confirm in considerable degree the observations of Lowell."[54] Significantly, several years later—and now further removed from the influence of Lowell's magnetic personality—he summarized these observations rather differently:

> I have studied Venus carefully at Flagstaff with the Lowell refractor and found it extraordinarily difficult to make out anything, so that I personally should be very cautious in relying upon visual observations. It is an old story . . . no doubt, and one I have amply confirmed, that with a small telescope one often is sure of large bright spots and faint shadings—which vanish from sight when the planet is studied with a large aperture in better air. Indeed my experience is that the better the conditions, atmospheric and instrumental, the less I can see.[55]

The criticism of Lowell's Venus observations marks, in certain respects, the beginning of his *peripeteia,* or tragic reversal: the pleasantly hallucinatory state, not unlike that of Coleridge dreaming his vision of Xanadu, is now threatened from without by the intrusion of what Freud called the "reality principle" and that Coleridge himself personified in his "man on business from Porlock." In an earlier chapter, Lowell's observatory was compared, perhaps somewhat fancifully, to the pleasure dome Kubla Khan decreed for himself. The observatory was seen as a psychological haven—a later version, perhaps, of the house that Lowell had found so pleasant in Seoul, where from the street one entered a courtyard, then a garden, "and so on, wall after wall, until you have left the outside world far behind and are in a labyrinth of your own."[56] Lowell had indeed taken his desire to leave the world behind to extremes. He had left the world behind altogether, in favor of another one among all the "wonderworld of stars," as he had written in his poem "Mars":

> Till as I peer, myself I see
> Sailing some landlocked Martial sea,

Scaling some Martial mountain crown,
To gaze from its vantage summit down
On a landscape whose strange mien,
Ruddy red where we look for green,
Pricks perception to grow more keen;
Or better still in some strange-built town
To mingle with the Martial men,
Enlarging my Earth-restricted ken,
Amid their pageant-seeming show
Learning a little of what they know,
Grown by my travel in some wise then
Myself a cosmic denizen.

But it is all a dream, of course, and he wakens from it:

Yet still I sit in my silent dome,
Wharf of this my island home,
Whence only thought may take passage to
That other island across the blue,
Against hope hoping that mankind may
In time invent some possible way
To that far bourne that while I gaze
Through the heaven's heaving haze
Seems in its shimmer to nod my way.

Professor W. Jackson Bate has said with reference to Coleridge's poem: "In this house of pleasure, where he hopes to be shut off from 'tumult and care,' his dreams are troubled by the thought of 'deluge and invasion' in his dominions. . . . In the very commitment of the imagination to its dream, the closed paradise for which it has hoped proves incomplete and ultimately threatened."[57] The "possible way" to get closer to the planet that Lowell had "hoped against hope" would be found was, in fact, already in existence. Larger telescopes than Lowell's own would bring Mars thus closer, but what they would show, Lowell would refuse to accept to the end of his life.

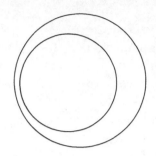

## CHAPTER 15

# *Apostates and Critics*

> *knife and glass*
> *That make thought physical, and*
> *thrust far off*
> *The Heaven, so neighborly with man*
> *of old,*
> *To voids sparse-sown with alienated*
> *stars,*
>
> . . . . . . . . . . . .
>
> *At best resolving some new nebula,*
> *Or blurring some fixed-star of hope*
> *to mist.*
>
> —James Russell Lowell,
> "The Cathedral"

PERHAPS THE most notable scientific contribution thus far of young Andrew Ellicott Douglass, who took over the direction of the Lowell Observatory during Lowell's illness, had been his investigation into the atmospheric basis of astronomical seeing—the second principal objective, after the study of Mars itself, that Lowell had outlined for his new observatory. His results were published in a longish paper that appeared in *Popular Astronomy* in June 1897 under the title "Atmosphere, Telescope, and Observer." In this paper Douglass discusses atmosphere in its relation to astronomical observation in essentially the correct frame of reference. Moreover, what he says is of importance for the later polemical use made of his results by Lowell, who always maintained, against E. E. Barnard and others, that modest instruments were more useful than large ones for planetary work. This very point is, in fact, Douglass's opening volley, for he

notes that

it is a matter of importance and significance that so little has been done in recent years upon planetary detail by telescopes of great size. The strenuous effort to produce instruments of enormous power and perfection has resulted in telescopes of remarkable light-giving capacity . . . but which do not show improved definition. There is no difficulty at all in assigning poor atmosphere as the cause of this.[1]

Douglass goes on to emphasize that every possessor of a fair-sized telescope may study for himself the atmospheric conditions accompanying good and bad seeing simply by pointing the telescope at any bright star, taking the eyepiece out and placing the eye in the focus of the objective, then watching the streams of air passing overhead (these appearing as a series of bright and dark waves, or "shadow bands," moving across the field). From his own experiences at Flagstaff, he had found that "when the seeing was good the currents seen through the telescope . . . came from the east," but when the seeing was bad, "they came from the north or northeast and the mountain range extending from ten miles due north to about six miles east-northeast was shown to be responsible."[2] Thus, he concluded, "it seems probable . . . that neighboring mountain ranges are not good"—the first published hint that the seeing at Flagstaff might not be as exceptional as officially claimed.

The often-debated question of the relative usefulness of large and small telescopes in planetary observation is found to depend on the average wavelength of the atmospheric "waves" (or equivalently, the diameter of the cells of turbulence) passing overhead. If the wavelength is greater than the aperture of the objective, the image moves back and forth with little blurring of detail; if it is less than the aperture, the image will appear steady in bodily position, but detail will be confused. Moreover, there is usually more than one set of atmospheric waves present, and they may all be of different sizes:

By a big diaphragm we can get rid of the blurring effect of the largest set. By medium and small diaphragms we can improve successively the bad effect of the other series, but in doing so the light is enormously decreased. We may summarize this matter of aperture by saying that the smaller the aperture the more bodily motion and less confusion of detail; the larger the aperture, the less bodily motion and the more confusion of detail. This leads us directly to the aperture required for certain classes of work. For seeing planetary detail we should use a small aperture unless the seeing is at its very best.[3]

In a word, the aperture must be matched to the seeing. The diaphragm, which Barnard did not find helpful in his observations with the 36-inch refractor on Mt. Hamilton, was henceforth to be for Lowell and his staff arguably the most indispensable piece of equipment in the planetary observer's armamentarium—the great improvement in definition obtained with stops being underscored time and time again in Lowell's own pronouncements on the subject.

In fact, Lowell seems to have been quite right with regard to his own telescope. Planetary images in his 24-inch do seem to be improved when the full aperture is stopped down. Carl Lundin of the Clark firm, who was at Flagstaff in 1909 for the installation of Lowell's new 40-inch reflector, told Lowell: "Before I came here I did not believe what you said about diaphragming down and getting better images, but I see it is so."[4] For what it's worth, I can add my own testimony to this effect.

Yet the proverb "Beware the man of one book" is no doubt as true of telescopes as of books, and Lowell had no right to suppose that other observers using different telescopes at other sites, would find precisely the same thing. He declared in 1910, vis-à-vis Antoniadi, who had used the full aperture of the Meudon 32¾-inch to observe Mars the previous year: "We find that even in our air, which is selected for the purpose, and is much better than at Meudon, we can rarely use 24 inches to advantage. With us the maximum efficiency lies anywhere from 18 to 12 inches."[5] Yet as we have seen, the reputed excellence of conditions at Flagstaff was actually a matter of considerable private dispute. Conditions at Meudon, ironically, may have been less disadvantageous than Lowell had supposed; smog over a city can have a steadying effect on planetary images. Finally, it is not clear that atmosphere was the only or even the primary reason for the diaphragm's usefulness with the Lowell 24-inch; Clyde Tombaugh has claimed, as we have

seen, that because of the telescope's relatively short focal ratio,
chromatic aberration causes planetary images to deteriorate
when more than 20 inches of the full aperture are used.[6] The
diaphragm increases the effective focal ratio, thus decreasing
the separation of the red and blue focuses, though at the cost,
of course, of resolving power.

As Lowell's observations came increasingly under attack,
the atmospheric argument became more and more vital to the
defense. As early as 1897 he had alleged that the reason that
other telescopes had failed to show his Venus markings was
simply that their air was not good enough. In particular he
argued that Barnard's failure at Yerkes was hardly surprising,
given the poor atmospheric conditions that were generally to
be found there.[7]

Lowell overlooked the fact that Barnard had also failed to
see the markings at Lick. Barnard himself had an open mind
on the matter and actually wanted to go to Flagstaff to see for
himself. He wrote to Douglass:

> I am immensely interested in this matter. I do not have
> any ill feelings—why should I have? It is simply to satisfy
> my own eyesight that I want to come. Dr. See and Mr.
> Lowell have always been friendly towards me and if for
> nothing else, I should want to come so that I shall cease
> to do them an injustice, for I doubt not that my failure at
> Lick Observatory to see these things so easily seen from
> Flagstaff may have had some sort of influence against the
> observations.[8]

Barnard did go to Flagstaff in 1898, but there were such
strong winds during the visit that no satisfactory observations
could be made; one can only suppose that he returned to
Yerkes rather unimpressed. One would very much like to
know what Douglass and he discussed, for by this time Douglass had himself begun to grow skeptical of Flagstaff observations—including his own. By December 1898 he had commenced a series of experiments in which he attempted to observe artificial planets through the telescope from about a mile
away and found "at once that some well known planetary
appearances could, in part at least, be regarded as very doubtful, such as Bright Limb and white cusps"—references to

various telescopic aspects of Venus—"and occasional indefinite markings."9

Douglass had already done experiments with artificial planets during his days with Pickering in Peru, and he had discussed doing more with Lowell himself as early as 1895. At first Lowell was fairly receptive to the "psychological question" in planetary observation. In his book *Mars,* published in December 1895, he acknowledged that it was possible for stimuli

> to be started in some other part of the brain, travel down to the lower centres and be sent up from them to the higher ones, indistinguishable from *bona fide* messages from without. Bright points in the sky or a blow on the head will equally cause one to see stars. In the first case the eyes were duly affected from without; in the second, the nerves were tapped to the same effect in mid-route; but in each case the subsequent current travels to the higher centres apparently as authentic the one as the other.[10]

At this time Lowell, not long back from the Far East, where he had studied psychic phenomena in the form of Shinto trances, seems to have been intrigued by what he describes in a letter to Douglass on January 5, 1895, as "optico-psychic" phenomena.[11] Following the sharp criticisms of his Venus observations, his attitude changed, and he became increasingly querulous about such psychological inquiries, ordering Douglass to break off his experiments in 1899. However, the following year he had a change of heart and allowed Douglass to continue; in particular, Lowell wanted to know whether the work shed any light on the Venus observations. He was then trying to write, as his vigor permitted, a memoir on Venus for the American Academy of Sciences. Otherwise, his main pastime was taking walks by himself, of which he wrote to his father: "I can converse with plants because they don't talk back."[12]

Douglass's experiments were still underway when Lowell finally returned to Flagstaff in the spring of 1901. Ironically, when Lowell himself tried his hand at drawing an artificial planet, he recorded a double canal where in reality there had been only a broad shading. By now, unbeknownst to Lowell,

Douglass had written to Joseph Jastrow, professor of psychology at the University of Wisconsin, summarizing his artificial planet work and noting that he would have written long before "but for Mr. Lowell's indifference to taking up the psychological question involved in astronomical work." Douglass proposed the following areas for investigation:

1. Find out if any definite forms exist in eye or from diffraction on white circular or gibbous disks . . . seen against a black (or purple) background.
2. Imitation of one's own drawings or those of others; what role does that play?
3. Geometric interpretation of difficult detail; how much does this affect the result?[13]

At this time he also penned the note in which the remark mentioned in the last chapter about Lowell's "hunting up a few facts in support of some speculation" appeared—and improvidently sent it to Lowell's brother-in-law, William Lowell Putnam, who had served as trustee of the observatory during Lowell's illness. Lowell dismissed him soon afterward, and though Douglass always maintained that he had been treated unfairly, further association between the two men was clearly out of the question.

Yet Lowell himself had vacillated on the reality of the markings on Venus. By 1902 he had gone so far as to publish a retraction—just about the only occasion during his astronomical career in which he retracted anything. In the journal *Astronomische Nachrichten,* the same in which his observations had first been published, he admitted that the center of his system of spokes shifted, not in accordance with the rotation of the planet, but so as to remain in the center of the disk as the phase increased—the circumstance that had prompted Antoniadi's scathing comment that the planet was thus proved to hold the same face to the Earth rather than to the Sun. He also acknowledged that experiments with artificial disks had shown similar markings under corresponding conditions, though no such markings really existed on the disks themselves, and speculated that these markings "might be caused by the eye wandering quickly from one of the dark indentations to the centre, and thus dulling unconsciously a path along the retinal rods."[14] The generous Maunder, instead of gloating, encouraged the members of the British Astronomical Association during its meeting of October 29, 1902, to applaud "the honourable and candid way in which Mr. Lowell has been so prompt to express a doubt as to the reality of his own supposed discoveries."[15]

Unfortunately, when Lowell returned to the telescope again the following year he once again found the same markings staring back at him. This time he made every effort to guard against "self-deception":

> Experiments on the visibility of a wire show that it is possible by direct consciousness to part the true from the spurious. Although it is possible to see illusory lines, it is also possible to be cognizant of the fact. If one pay attention, an hallucination of the sort may be found to differ from a presentation of fact by the absence of the sense of reality. I am speaking, of course, only of one kind of hallucination, upon which I have myself made experiment. A peculiar consciousness of objectivity accompanies an impression started from without which is wanting in one originated from within.[16]

Attempting to apply these criteria to the Venus markings, he noted that they "came out at times with a definiteness to convince the beholder of an objectiveness beyond the possibility of illusion," and cited in support of this contention several entries from his observing log:

> March 22. Never felt more sure of the reality of the markings.
> April 13. These markings are as sure as markings can be.
> May 22. The markings came out like an etching. No illusion effect whatever.[17]

Though Lowell had thus persuaded himself once again of the reality of what he was seeing on Venus, he expended relatively little effort in publicly defending them, for by 1903 a formidable challenge had begun to mount concerning the Martian canals themselves.

THE SAME YEAR, 1903, in which Lowell "rediscovered" his markings on Venus saw the publication of an important paper by Maunder and J. E. Evans entitled "Experiments as to the Actuality of the 'Canals' of Mars." It was an elaboration of the idea that Maunder had put forward in 1894, that the canals of Mars might actually be due to "the summation of a complexity of detail far too minute to be ever separately discerned." Boys at the Royal Hospital School in Greenwich had been asked to reproduce a disk they were shown on which no canals had been drawn but only "minute dot-like markings." Maunder and Evans found that when the disk was viewed from a certain distance, the boys drew "canals":

> So far as the experiments go, the most fruitful source of the canal-like impression is the tendency to join together minute dot-like markings. If these are fairly near to each other it is not necessary, in order to produce the canal effect, that they should be individually large enough to be seen.
>
> But it is not necessary that the markings should be dot-like, that is to say, approximately circular. They may be of any conceivable forms provided only that they are outside the limit of distinct vision and are sufficiently sparsely scattered. In this case the eye inevitably sums up the details which it cannot resolve into fine lines essentially "canal-like" in character.[18]

Flammarion, on attempting to repeat Maunder's experiment with French schoolboys, found that they did not show canals. Lowell was equally unimpressed by what he called the "small boy theory" and argued in a letter to Maunder:

> First, because that *a* may produce the effect of *b* furnishes no proof or even presumption that *a* is *b* since *b* itself would produce the same effect. Secondly, the observations at Flagstaff are of a much greater definiteness than is supposed in England. You might, for instance, as well say that a telegraph line as seen from a street was merely an optical effect of dots and shadings. This is a matter of astronomical experiment and not of psychologic discussion—the eye may be deceived below a certain limit but not above it.

The question was finally to be decided, he insisted, by "actual observation directed to that end," and striking a familiar note expressed confidence that "if England would only send out an expedition to steady air with acute-sighted, not sensitive observers . . . it would soon convince itself of these realities."[19]

Lowell was undoubtedly right in insisting that the question was not to be settled by having recourse to experiments with artificial planet disks but only by direct observation at the telescope. Maunder had shown that dots and shadings *might* give rise to an impression of continuous lines, but of course, as Lowell was right to point out, continuous lines would also give that impression! Nevertheless, by 1903 a few skilled observers had begun to confirm, in actual work at the telescope, that the surface of Mars did seem to be composed of complex structures lying generally outside the limit of distinct vision, as Maunder had suggested. An Italian astronomer, Vincenzo Cerulli, had independently arrived at a view similar to Maunder's based on insights like the following, obtained in 1897 from his observatory at Teramo with a 15½-inch refractor: "The night of January 4 yielded some moments of perfect definition [in which] Mars appeared perfectly free from undulation. . . . It seemed to me that [the canal] *Lethes* was losing its form of a line and altering itself into a complex and inextricable system of . . . very small spots."[20] The canal that appeared thus resolved to Cerulli's eye was, by the way, none other than that which had been the first to declare itself to Lowell and Pickering at Flagstaff in 1894.

Cerulli had himself been one of the most prolific observers of canals. Another observer who had recorded numerous canals but who later came to have second thoughts was the British amateur Percy B. Molesworth, who in 1903 wrote: "The amount of detail is bewildering, and I despair of giving even an approximate idea of it in a drawing." Again, "the broad effects one draws are simply the combined results of myriads of small details, too minute to be appreciated separately." And, "I cannot help being certain that our present instruments are quite incapable of dealing with the details of Mars, and that even the best and most careful drawings give an utterly wrong idea of the configuration of his surface. The eye interprets as well as it can, but the task is beyond its power."[21]

The visual observers could not, then, quite agree as to the true shape of the Martian surface, and though Lowell could contend that "it is possible by direct consciousness to part the true from the spurious," this consciousness was necessarily in the domain of the subjective and could not be imparted to another. The affidavits that Lowell had his friends write concerning what they had seen through his telescope were, similarly, of as little avail in convincing those who did not see. What was needed was objective—indisputably objective—evidence. By 1905 it seemed that Lowell had actually managed to produce it.

Since 1901, Lowell had had various assistants working on the project of photographing Mars with the object of thus recording the canals. But two difficulties had stood in the way. One was "the varying airwaves which now favor, now prevent, the definition of such fine detail as that of the canals." The other was "the insufficient speed of the photographic plate":

> In the registering of such detail the eye has a great advantage over the camera; for it can perceive much more sensitively [acutely?] than the plate and furthermore retains an image only for the twentieth part of a second. It can thus record a moment of apparition; the camera cannot, but must take the good with the bad and yield only a blurred composite picture of both.[22]

Despite these overwhelming difficulties, by 1905 Lowell's assistant, C. O. Lampland, was widely reported to have actually succeeded in photographing some of the canals, a feat for which he received the 1907 medal of the Royal Photographic Society. For their day, the photographs certainly were remarkable. Lowell took pleasure in pointing out that again "the essential factor that brought success was the one which has been found so vital to visual observation—the diaphragming down of the objective to suit the atmospheric current at the time of the observation. Not only did a diaphragm prove better than the full objective but the increased gain in definition was so great as to much more than offset all the bad effects of prolonged exposure" (in the case of Lampland's images, about eight seconds on average). Lowell called attention to the fact that, under the circumstances, one had no right to expect perfect clarity: "Inasmuch as such fine detail as the canals, owing to the airwaves, play bo-peep with either observer or camera, it is not to be expected that the more delicate of them should appear in every print. Yet they turn out to come nearer to doing so than could have been anticipated."[23]

Though Lowell, in his 1906 book *Mars and Its Canals*, boasted that "thus did the canals at last speak for their own reality themselves,"[24] it proved difficult to reproduce the tiny, delicate images, each only a quarter inch across. The reader whose interest had been piqued by Lowell's comments would thus have been disappointed to find that none of his photographs actually appeared in the book. While a number of the experts who had the opportunity to examine actual prints of the photographs concurred with Lowell in seeing canals in them, an equal number failed to do so. More important, even where it was admitted that there seemed to be linear details in the photographs, it did not necessarily follow that the photographs proved that this was the actual form of the true features of the Martian surface, any more than the testimonials of visual telescopic observers to the same effect had done. As W. T. Lynn noted in an article in the *Journal of the British Astronomical Association*:

> The conclusions arrived at by Mr. Lowell are not quite so decisive of the continuity of these remarkable objects as may at first sight appear. Those who have seen the prints differ in their interpretation of the appearances; and Mr. Wesley [a leading expert on the interpretation of astronomical photographs] remarks that "doubtless a photograph has no imagination, but imperfect definition applies equally to photographic and visual observations. That which is seen as photographed imperfectly as a smooth continuous line *may* be full of small irregularities, and may not be strictly continuous.[25]

Images of Mars obtained by another of Lowell's assistants, E. C. Slipher, with an 18-inch refractor borrowed from Amherst College and set up at Alianza, Chile, during the 1907 opposition of Mars were—though significantly better than Lampland's pioneering efforts—no more decisive about the canals, nor was the 40-inch reflector that Lowell hoped to turn toward the planet in 1909 in order to combat the allegation that no large telescope had ever shown the canals. Though a few observations were made with the new telescope in this and subsequent years, it never did perform up to Lowell's expectations, probably because the telescope was handicapped by bad seeing resulting from the misguided idea of mounting it below ground level. Though it did show a few canals on occasion, it showed them less perfectly than the 24-inch refractor and succeeded only in reinforcing Lowell's prejudice against such large instruments for planetary work.

LOWELL WAS NOT the only one to turn a large telescope on Mars in 1909; another was the skilled observer Eugène Marie Antoniadi, who used the great 32¾-inch refractor at Meudon, between Paris and Versailles, that year for some of the most remarkable views of the planet obtained in the pre-spacecraft era.

Though his name has come up often in the preceding pages, Antoniadi has yet to be properly introduced. Of Greek descent, he was born at Constantinople (now Istanbul) in 1870. He began observing with smallish telescopes at Constantinople and on the island of Prinkipo in the Sea of Marmara in his late teens, and some of his observations were published in Camille Flammarion's *Bulletin*. He visited Flammarion at Juvisy in 1893 and stayed on as his assistant until 1902. After his departure from Juvisy he remained in France for the remainder of his life, becoming a naturalized citizen and receiving its highest honor, the Cross of Chevalier of the Legion of Honor. He died in 1944.[26]

One might wonder about the association of Antoniadi—later one of the severest critics of Lowell and the canals of Mars—with someone having views as extreme as Flammarion. But at first they were quite compatible in their visions, together producing several canal-filled maps of the planet with Flammarion's 9½-inch refractor. Indeed, it may be worthwhile, before discussing Antoniadi's famous observations with the Meudon refractor in 1909, to indicate the gradual development of his skepticism toward the canals, which had already become fairly complete even before he observed with the large telescope.

Between 1894 and 1916, Antoniadi was director of the

Mars section of the British Astronomical Association. As such, he was responsible for drawing up its regular reports. As one reads through his successive publications, one sees a gradual change in his views—with, perhaps not surprisingly, a particularly significant development taking place after he left Flammarion in 1902. In his report on the observations of Mars of 1896–97 he notes that "the canals were seen by all the working members of the section invariably."[27] Indeed, Antoniadi himself, with forty-six sighted, was third behind only Percy Molesworth and Rev. T.E.R. Phillips in the number recorded. Phillips, observing for the first time with a 9½-inch reflector, went so far as to say: "My experience of Martian observation this winter has led me to believe that Mars is not nearly so difficult an object as is commonly supposed, and that many of the canals are easy." Antoniadi was not quite so sanguine, declaring that in his experience at Juvisy "the canals are very difficult objects, visible only by rare glimpses"; and he added that "but for Prof. Schiaparelli's wonderful discoveries, and the foreknowledge that 'the canals were there,'" he would have missed at least three-quarters of those he did see. Yet despite his willingness to accept the existence of the canals *in general,* Antoniadi nevertheless felt it prudent to set aside the work of the Scottish amateur C. Roberts, which two years before he had praised enthusiastically. Roberts, who along with Spiridion Gopcevic (or Leo Brenner, as he styled himself) seems to have been one of the most wildly imaginative—or suggestible—of the observers of the day, had reported seeing no less than 134 canals with a 6½-inch reflector, but Antoniadi wrote: "It was thought safer to avoid introducing uncertain data in the general excellence of the section's work, and not overcrowd our already crowded chart with the most daedelian canal network ever devised."

Antoniadi's report on the results of 1898–99 included the following comment:

> Notwithstanding the natural skepticism of many scientific men, every opposition brings with it its own contingent of confirmations of Schiaparelli's discovery of linear markings, apparently furrowing the surface of the planet Mars. The difference between objective and subjective in the daedelian phenomena presented by these appearances

will be the work of future generations. But the value of
the great Italian results will be everlasting.[28]

The discovery by Lowell and his assistants of what Antoniadi in 1903 referred to as "subjective" linear markings on
Mercury, Venus, and the Jovian satellites, seems to have
played a significant role in shaking his confidence in the Martian canals as well. Lowell, of course, often insisted that the
markings were quite unlike one another. He wrote of the
markings he had observed on Venus, for instance, that

> it appears necessary, in view of the widespread and per-
> sistent misunderstanding of what the markings are, to
> assert again that they are in no sense canaliform. They
> bear no resemblance whatever to the "canals" and oases.
> They are not of even width, are not dark and sharp cut
> and do not form a system of interlacing lines. Nor are
> they ever double. . . . Arguments applied to the one set
> are quite inapplicable to the other. Furthermore, they are
> of a much higher order of difficulty. Unless the condi-
> tions of visibility are such as to show an observer the
> "canals" of Mars with ease and certainty it were useless to
> attempt this much harder planet.[29]

Yet Lowell's sketches spoke much more eloquently for themselves—and to quite opposite effect. A quote from Lowell
himself, intended to prove a different point, serves as an
ironic comment here. Writing in 1905 of the unwillingness of
"conservatives" to accept new ideas, he brought up the Owen-
Huxley controversy of 1860–62 "in which Owen would not
have it that the ape's brain bore similarity to man's and when
facts opposed him loftily ignored them."[30] The applicability
of this to the various planetary markings that he sketched
needs no further comment, it being sufficient merely to mention that their similarity was likewise loftily ignored by
Lowell.

Rather than being the work of future generations, as Antoniadi had previously thought, by 1903 the business of dif-
ferentiating the real from the spurious in planetary markings
had become only too urgent. Antoniadi did not yet go so far
as to put the Martian canals in the same category as the other
Lowellian markings, pointing out that "the hard line-likeness

of the 'canals' is almost sure to be experienced by painstaking observers of the planet; and this circumstance cannot be treated lightly as illusive."[31] Nevertheless, in order to avoid the "possibility of our representation of Mars [being] profaned by doubt," he took a decisive step. In addition to the usual canal-filled chart, which was prepared on the basis of the submitted observations, he published another from which all the canals had been carefully excluded. Maunder wrote of this chart:

> We seem to have returned to the pre-Schiaparellian age. . . . Is it a retrogression or an advance? . . . Either way we may take it as marking an epoch; for it is practically the first time for five-and-twenty years that a chart of Mars has appeared in which the canal-system was not predominant. Even should it be condemned as unscientific, it would still have an historic importance as marking the growing strength of a reaction.[32]

Of course Maunder never thought for a moment that it would be this chart that would be declared unscientific, but rather all the others that had appeared in succession ever since 1877, when the canals had first become fashionable. But though admitting that it was only too probable that the canals would have to be given up, he added magnanimously:

> We need not for one moment make the assumption that the various observers of the canals have not actually seen what they have represented, or that they have badly represented what they have seen. The question is one simply of the limitations of our sight. Under certain conditions the impression as of lines is given to us by objects not linear—an impression in no way differing, so far as we have yet ascertained, from that which actual lines would have produced. . . .
>
> Nor will the entire abolition of the canal-system as such efface the debt which we owe to Schiaparelli for having brought it into notice. For the time it has been the nearest approach we could make to a true perception of the surfacing of the planet, and if it has brought into prominence scarcely suspected sources of error, we are now gainers by having had these dangers fully exposed.[33]

Ever the gentleman, Maunder thus closed with kind words even for his scientific opponents. By 1903, Antoniadi was definitely leaning toward his camp, and indeed might already have considered himself a member of the "growing reaction." A man with a mind as incisive as Maunder's, but with far more relish for controversy, a skillful writer in English as well as in French (the more remarkable in that neither of them was his native language), Antoniadi had, moreover, a genuine gift for biting sarcasm. In later years he would become the leading critic of the canals—easily a match for Lowell in every respect —and in particular would argue in a series of articles based on his own remarkably detailed observations of 1909 that with them the canals had been totally disproved.

BETWEEN SEPTEMBER 20 and November 27, 1909, Antoniadi observed Mars with the 32¾-inch refractor of the Meudon Observatory, at the kind invitation of the observatory's director, H. Deslandres. The great lens, the masterpiece of Paul and Prospero Henry, was completed in 1889, and the telescope became operational two years later. It was, when Antoniadi observed with it, the largest of its type in Europe; indeed, it has not been surpassed to this day.

Shortly before commencing work with the great refractor, Antoniadi resumed a long-dormant correspondence with Lowell, apologizing for the manner, if not for the fact, of his criticisms of Lowell's Venus work and admitting that this had been "somewhat sharp and displaced." In addition, he asked for some of Lowell's recent publications and suggested that in return he would like to do something for Lowell: "I draw well, and if I can be of any use to you in that line, I hope you shall not hesitate to apply me in need."[34] Antoniadi could indeed draw well; he was perhaps the most skillful draftsman ever to apply himself to the exacting art of planetary portraiture. Lowell in his reply failed to mention whether he wished to avail himself of Antoniadi's artistry, but he did take occasion to offer his usual advice: "I am glad that you are to use the Meudon refractor but remember that you will have to diaphragm it down to get the finest details. Even here we find 12 to 18 inches the best sizes."[35]

By the time Lowell wrote this, Antoniadi had already begun work with the great refractor. It is interesting, in light of some criticism that has recently appeared concerning the supposedly low powers used by Lowell with the 24-inch refractor—usually 310× or 400×[36]—that Antoniadi also preferred to use relatively low powers, noting that with 320× he saw "faint half-tones never suspected before," while 470× and 540× "were already too much for the delicate shadings."[37] He also noted, significantly, that with the great telescope there was "no spherical or chromatic aberration troubling the image."[38]

In the account of his observations that he sent to Lowell, Antoniadi readily admitted that "here, in N. France, we seldom have really good images," and he expressed envy at the "ideal definition" Lowell and his colleagues presumably enjoyed at Flagstaff. Nevertheless, despite Lowell's urgings to the contrary, Antoniadi used the full aperture of the telescope, and in fact he was quite incredulous that Lowell could believe that the 24-inch, stopped down to 12 or 18 inches, could equal in resolving power the 32¾-inch. Antoniadi later sent Lowell several of the drawings he had made with the great telescope (Fig. 15.1) with the comment that "my Meudon work has surpassed all my expectations," and he pointed out that "the tremendous difficulty was not to *see* the detail, but accurately to *represent* it. There, my experience in drawing proved of immense assistance."[39] The detail was not, however, geometric but instead highly irregular and of completely natural appearance. "Bewildering" was the word he used again and again to describe it.

Lowell rather perversely dubbed as the best the one drawing Antoniadi had marked "tremulous definition." This one, needless to say, showed canals. The others, which had shown the planet as it had looked in "moderate," "splendid," and "glorious" seeing, he dismissed as "not so well defined," and he lectured Antoniadi once again on the problems with seeing with large telescopes:

> This is the great danger with a large aperture—a seeming superbness of image when in fact there is a fine imperceptible blurring which transforms the detail really continuous into apparent patches. On the other hand, a bodily movement often coincides with the revelation of fine de-

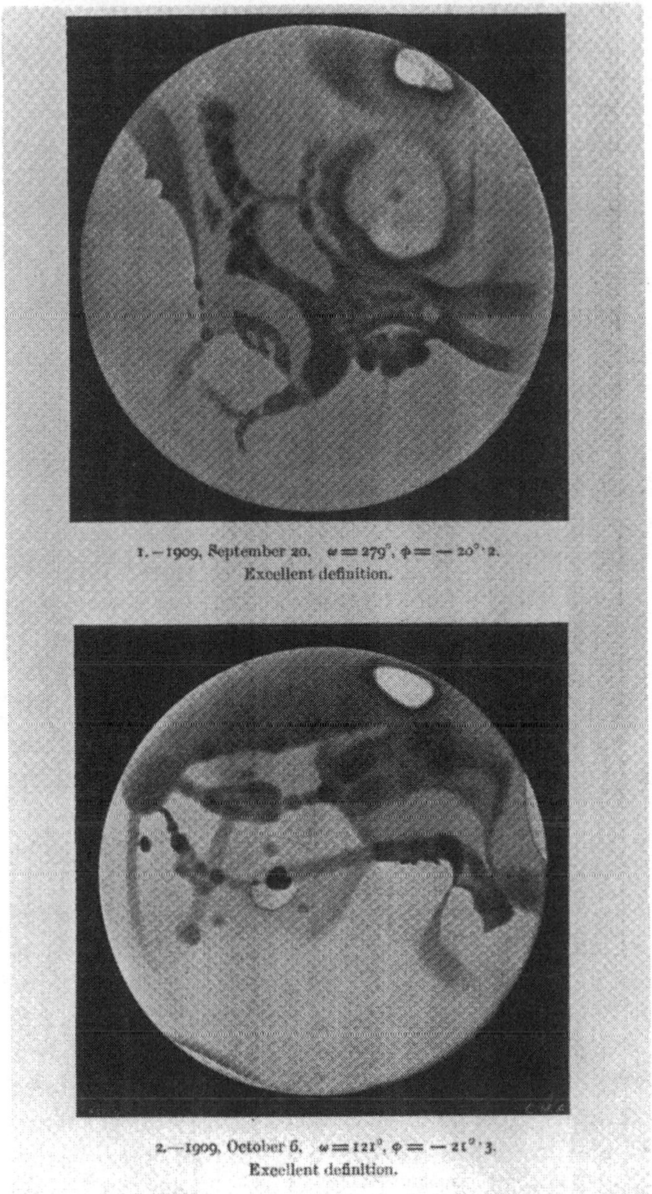

1. — 1909, September 20. $\omega = 279°$, $\phi = -20°.2$.
Excellent definition.

2. — 1909, October 6. $\omega = 121°$, $\phi = -21°.3$.
Excellent definition.

*Figure 15.1*
*Drawings of Mars by E. M. Antoniadi,
September 20 and October 6, 1909,
two of those he sent to Percival
Lowell. The September 20 drawing
shows the region centered on Syrtis
Major and was made on Antoniadi's*
*first night of observing with the
great Meudon refractor, when he
enjoyed several hours of exceptional
seeing. Note the "leopard skin"
appearance of the Mare Tyrrhenum
in this drawing.*

tail. This subject we have carefully investigated here and all of our observers recognize it.[40]

Antoniadi wrote back, with understandable irritation at Lowell's rather condescending tone: "I understand from your letter that you consider my knotted Mare Tyrrhenum as due to blurring; but I beg to call your attention to the fact that I was holding steadily this knotted structure; and that two days ago, I found this particular knotted appearance confirmed by photography."[41] It was this region which Antoniadi was later to describe, in a vivid and particularly apt phrase, as looking "like a leopard skin."[42]

To Antoniadi's mind, Lowell's quibbles had all the force of a pygmy's lance. Far more compelling than Lowell's arguments was what Antoniadi had seen with his own eyes at the great telescope, and a profound and unforgettable experience it had been. With great skill and thoroughness, Antoniadi recounted in a series of publications the details of what he had seen, particularly on the evening of September 20, when the air permitted, for seven hours, images that were nothing less than glorious. He became henceforth the most zealous apostle of large telescopes—and the most formidable critic of the canals seen in smaller ones. The Martian "deserts" had appeared "covered with a maze of knotted, irregular, chequered streaks and spots," as he wrote to Lowell. In the desert known as Amazonis he had held some of this detail for 10 to 12 consecutive seconds, compared with the brief glimpses under which, in terrible seeing, "hideous lines" had been wont to appear. Even held thus steadily, the details were so intricate that they could not be drawn, but for Lowell's benefit Antoniadi provided an impressionistic sketch showing approximately what they had looked like (Fig. 15.2).[43] As mentioned above, on the night of September 20 the steady images lasted for hours, and Antoniadi completed a beautiful and remarkably detailed drawing of the Syrtis Major region. "After my excitement at the bewildering amount of detail visible was over," he wrote to Lowell, "I sat down and drew correctly both with regard to form and intensity all the markings visible. . . . However, one third of the minute features I could not draw; the task being beyond my means."[44]

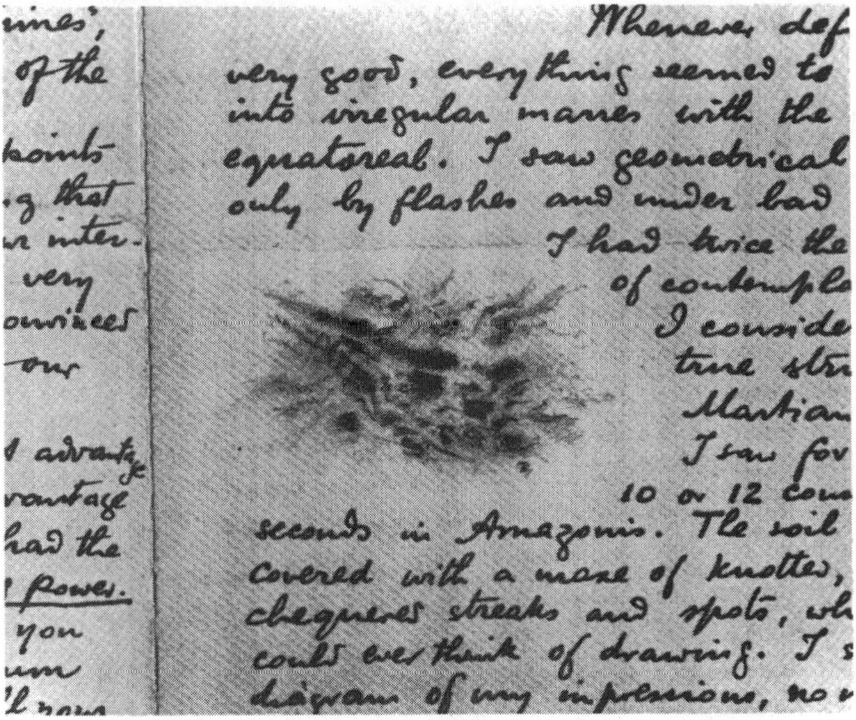

Whenever def
very good, everything seemed to
into irregular manes with the
equatoreal. I saw geometrical
only by flashes and under bad
I had twice the
of contempla
I conside
true str
Martian
I saw for
10 or 12 con
seconds in Amazonis. The soil
covered with a maze of knotted,
chequered streaks and spots, wh
could ever think of drawing. I s
diagram of my impressions, no

*Figure 15.2*
*Antoniadi's drawing of the desert
area in Amazonis, from a letter to
Percival Lowell dated November
15, 1909.*

The best seeing at Meudon was on nights when it was "foggy" and there was no wind. On most evenings, however, the images were "more or less boiling"; even so, Antoniadi was never tempted to have recourse to diaphragms, finding that "during the less agitated moments of a boiling image" the larger aperture revealed more markings than any smaller one could ever do.[45] The good views, he found, were "preceded by a period of slight rippling of the disk, very detrimental to the detection of fine detail. The undulations would then cease *suddenly,* when the perfectly calm image of Mars revealed a host of bewildering irregularities."[46]

These views helped to consolidate his ideas about the planet's surface. But it had been the superb images of the first night, September 20—doubly impressive for being his very first with the large telescope—that had come upon him with all the force of a revelation. Indeed, such in a real sense it was. As he recalled it afterward for the benefit of the members of the British Astronomical Association's Mars section:

> At the first glance cast through the 32¾-inch . . . the Director thought he was dreaming and scanning Mars from his outer satellite. The planet revealed a prodigious and bewildering amount of sharp or diffused natural, irregular, detail, all held steadily; and it was at once obvious that the geometrical network of single and double canals discovered by Schiaparelli was a gross illusion.[47]

He later felt the need to defend his priority in seeing what a large telescope could show on Mars, pointing out that his first report on the observations of September 20 had been published eight days before the famous telegram of E. B. Frost, who had answered a question about the success of the Yerkes refractor on the canals by wiring: "Yerkes telescope too powerful for canals."[48] In fact, as we have seen, Barnard had had similar impressions of Mars with the Lick 36-inch in 1894, though priority would actually seem to belong to Charles E. Burton, who, observing with the 3-foot Rosse reflector in 1869, had noted: "the quantity of details is very considerable, the most remarkable being the rather blackish, exceedingly minute points abundantly distributed over the length of the neck of the Hourglass Sea. This revelation was evidently a result of the aperture."[49] Antoniadi was not even the first to resolve Mars into a complex of irregular details with the Meudon instrument. Gaston Millochau, a former director of the observatory, had done so between 1899 and 1903, as Antoniadi later found out.[50] But certainly Antoniadi was far and away the most effective spokesman on the subject, and in any case a number of similar observations were made with other large telescopes during 1909, most notably by George Ellery Hale and a group at Mt. Wilson (including, ironically, Lowell's former assistant, Douglass), who used its 60-inch reflector diaphragmed to 44 inches to decipher on the planet so much intricate detail that they also had found it impossible to draw.[51]

Clearly, such observations demanded an answer from Lowell. As Antoniadi put the matter bluntly, "Prof. Lowell's canal network disappears under the worst conditions for it, that is, when much more delicate detail is quite plain."[52] Again, "Under good seeing, almost all Schiaparellian 'canals' visible either broke up into most complicated and irregular groups of shadings or became the mere indented edges of these shadings," usually the former.[53] In Antoniadi's view, Maunder had been completely vindicated. Lowell was not prepared to give up, though, and by January 1910 he had submitted a rebuttal of the most recent drawings of Mars to the British journal *Nature*. It had been atmospheric disturbances, Lowell argued, that had caused Antoniadi to see the surface features of Mars as irregular and all broken up. He later put his position quite succinctly:

> Every telescope makes a real image of a star, but owing . . . to the interference of the light-waves that image is not a point, but a circle with concentric rings of light around it. If you have a small aperture, from 2 to 6 inches, and the light is at all favorable, you will be able to see the pattern, and if you have good air it will come out beautifully as a sort of engraved etched thing. Now, as you increase the aperture, the image gets worse. Theoretically it should grow smaller, and your pattern remain the same; and so it does, but it also always grows worse. What first happens is that the rings begin to waver, and

they finally waver so much that they break out into a
mosaic of points. Then the main disc does the same, and
finally the whole thing is a beautiful spangled conception.
It is a mosaic from having been a series of lines.[54]

Something akin to this "beautiful spangled conception," then,
had been what Antoniadi had been observing on Mars with
the large refractor. The Martian details—in reality, lines—
had similarly broken up into a mosaic of points. (Antoniadi
was later to point out, however, that Cassini's division, seen
in front of the ball of Saturn, "is a line fully comparable with
the canals Prof. Lowell sees on Mars. Yet no large glass has
ever broken it up into an irregular mosaic, held steadily, and
the writer never saw it so sharply defined as with the Meudon
refractor.")[55]

Lowell's theory, indeed, seems to have been quite as spurious as the disks on which it was based. Any small telescope
will show a spurious disk and rings; as we have seen, the
smaller the better. To get the best view of the "real form" of
the spurious disk, one would be advised to use a lens of a half
inch or so. But of course the actual image of the star is, as
Lowell well knew, a point, not a disk; the minute disk and
rings formed by the large telescope comes closer to representing this than does the blown-up disk of the small telescope. It
is quite true that the delicate image formed by the large telescope will be readily torn apart by every vacillation of the air.
But the apparent steadiness of the small-telescope image to
which Lowell attaches such importance is strictly artifactual—
a product of the image's coarseness, not of perfect definition.
The individual points making it up are also in trembling motion, but, as Isaac Newton had written in the *Opticks* two
centuries before, "all these illuminated Points . . . [are] confusedly and insensibly mixed with one another by very short
and swift Tremors, and thereby cause the Star to appear
. . . without any trembling of the whole."[56] Thus the apparently steady image is actually a blur, and the whole discussion
is completely irrelevant to the question of the real structure of
planetary surface markings.

E. E. Barnard, for his part, was quite unimpressed by
Lowell's line of reasoning and wrote to Antoniadi on May 27,
1910:

Lowell seems to believe that large instruments cannot give definition good enough to show the geometrical canals. He forgets that the best proof of the power of definition of an objective is the separation of very close double stars. In this field, the work by Mr. Burnham and Professor Aitken with the 36-inch [at Lick] shows that measurements can be made of double stars which are so close that the Lowell refractor, under its atmospheric conditions, cannot hope to reveal them.[57]

Commenting on Antoniadi's published observations of Mars, Barnard understandably agreed "in respect to Mr. Maunder's work in trying to clear up the tangle about the canals of Mars."[58]

Perhaps still more relevant to the canal question than the resolution of double stars was the way that large and small telescopes performed on lines. Lowell, based on experiments in viewing telegraph wires, had found that the eye is able to detect lines when their width was only $1/100$th that of the smallest visible spot and declared: "Think what that means. Projected on to Mars it means that we could see a line one or two miles broad."[59]

In fact, however, a diffraction analysis shed a rather different light on the matter, as Antoniadi later pointed out. Starting with one of Schiaparelli's observations, made on June 9, 1890, in which the Italian astronomer had noted that the canal Jamuna had appeared to him to have a breadth of 0.04 seconds of arc, Antoniadi argued that this apparent breadth was actually that narrowed by diffraction. Given that the diameter of the Airy disk in the 18-inch refractor Schiaparelli had used on that occasion was 0".28, the actual breadth of the Jamuna (since the bright areas on either side of the stripe would encroach by half the diameter of the Airy disk) must have been 0".14 + 0".04 + 0".14 = 0".32. In the Meudon refractor, in which the diameter of the Airy disk is 0".16, the apparent breadth of the stripe would be 0".32 − 0".16 = 0".16, or four times that of Schiaparelli's line. Were the canals actually regular stripes on the surface of Mars, they ought, then, to appear as such in the large aperture, only broader. Such was the case, Antoniadi noted, with unquestionably true features of the sort—the Cassini division in Saturn's rings, for instance, as mentioned above. But, Antoniadi asked, "what is the aspect of the Jamuna in the great lens? Nothing like a regular band . . . but rather the mere irregular border of a weak halftone."[60] Such, in general, was the report he gave of the Schiaparellian canals in the great telescope. In the case of some of them he confirmed that there was an objective basis. "Schiaparelli's geometrical network of 'canals,'" he wrote, "appearing by flashes, is an optical illusion. In its place the globe of Mars shows either winding, irregular knotted streaks, or broad irregular bands, or groups of complex shadings, or isolated dusky spots, or jagged edges of half-tones."[61] It now seemed at last possible to pass judgment on the merits of the two sides in the debate between Green and Schiaparelli. "The verdict of science," wrote Antoniadi, "will probably be that, while the general appearance of the planet is certainly as drawn by the great English observer, yet the discovery of the 'canals' by the unrivalled acuteness of the Italian observer meant the discovery, under a symbolic, geometrical form, of the minor irregular shadings variegating the Martian surface."[62]

Antoniadi's admiration for the Italian astronomer was undiminished. If the canal network did not exist as Schiaparelli had depicted it, he wrote,

it is certain nevertheless that the "canals" of Schiaparelli have a basis in reality, so that in the positions of each of them, single or double, on the surface of the planet, there is present an irregular trail, a jagged edge of halftone, an isolated lake, in a word, something complex. These details are extremely varied. . . . Due to the exceptional acuity of his sight, Schiaparelli surpassed all the observers who worked with equally-sized instruments; but his modest refractors did not permit him a glimpse of the trains and other complex details in any other form than that of fleeting lines, for it would have been necessary for him to overcome the immutable barrier of diffraction in order to recognize the general irregularity of their appearance.[63]

Seen from the vantage point of three-quarters of a century, Antoniadi's arguments seem unimpeachable. Even at the time there was only one possible answer, and that was the one Lowell gave. In a surprise visit to the meeting of the British Astronomical Association in 1910, Lowell gave a brief talk before heading on to the Royal Astronomical Society to discuss his photographic work. In it he made the remarks, quoted above, about the way that atmospheric disturbance caused a diffraction pattern to break into a mosaic. Once again it had come down to the reduction of aperture to get the best images—the paradox Antoniadi later expressed thus: "As every reduction in the aperture is accompanied by a corresponding optical distancing of a celestial object, . . . it follows that in order to see a star one must distance it and render it indistinct."[64] Lowell's own summing up of his point was almost as paradoxically stated:

> It has been said that the best Indian of all is the dead Indian, and so the most perfect aperture you can use is no aperture at all. Just as you get down to this nothing, so your image will improve, because the air-waves which disturb are quantities of the first order to observers. . . . We find that even in our air, which is selected for the purpose, and is much better than that at Meudon, we can rarely use 24 inches to advantage, and as we diaphragm down we find improvement. With us the maximum efficiency lies anywhere from 18 to 12 inches.[65]

Fortunately for Lowell, it was the courteous Maunder, rather than the curt Antoniadi, who was sitting in the audience. Lowell showed some ignorance of the critical literature when he attributed to Simon Newcomb the idea that the lines on Mars were "a series of dots" (itself an oversimplification of the theory). Newcomb had indeed put forward this view—but several years after the English astronomer. Maunder did not bother to set Lowell straight on the matter of priority in his remarks at the conclusion of Lowell's address. Lowell had described his own experience in seeing Mars "like a steel engraving" and how well the canals had stood out in such definition. Referring to his experiments with telegraph wires, Lowell added that he had seen the canals "so well that I know the limit of visibility, and that limit is extremely small." Maunder, however, said that he deduced an entirely different conclusion from the experiments with telegraph wires, for since there was so wide a difference between the limit at which we have real definition of the object and the limit of mere perception, even though it be distinct, it followed that markings well within those limits—and Prof. Lowell had told them that the limit for mere perception was one-hundredth that of full definition—then those markings must tend to assume the simplest geometrical form—the forms, that is to say, of the circular dot and the straight line. He did not see that any of the observations which Prof. Lowell had laid before them had at all disturbed that fundamental principle.[66]

The argument concerning the canals had obviously become fairly technical, and at least some confessed themselves unable to pronounce between the two sides. M.E.J. Gheury wrote a short time after the meeting just described:

Now we are shown views of Mars . . . resembling some section seen through a microscope, and so absolutely unlike any telescopic views of Mars yet published as to be startling. These are explained by Prof. Lowell as due to the breaking up of the diffraction rings into a mosaic of points, so we have now subjective network versus illusive mottling. Prof. Lowell's wholesale disparagement of the performance of the largest instruments will, however, be received but with hesitation, as it is inconceivable that such magnificent equipment as some observatories do possess should be useless. Useless they are, if it be true that the images they give of objects are disfigured to such an extent that they are unable to show distinctly details that may be seen with refractors of moderate aperture, but instead exhibit purely optical diffraction patterns.[67]

Gheury pointed out that the matter was one of some seriousness, as the dispute was likely to spread to other planets as well—irregular details having frequently been reported on Jupiter which, he believed, were likely to be just as readily disposed of by Lowell's argument. In any case, life was too short to waste it on drawing "subjective reseaux or subjective mosaics." But when doctors disagree, who is to decide? Gheury recalled that a similar dilemma had existed in physics a few years earlier and had been settled when the disputants had come together in order to perform their experiments in the same laboratory. He suggested that Antoniadi and Lowell might do the same and that simultaneous work at one another's observatories might "advance the question more than if each spent years of the most persevering labour, engaged in deepening his own particular furrow, while the world looked on, unable to judge."[68]

This sensible idea was, unfortunately, never followed up. Lowell, despite his frequent visits to Europe, never seems to have expressed any interest in having a look through the Meudon refractor for himself. Doing so would certainly have been more to the point than his observations with his own 40-inch reflector, which appears never to have worked up to snuff. But suppose he *had* gone to Meudon, what would have been the result? Probably he would have come on a bad night and so would have left with his mind unchanged. But suppose he had come on a beautiful night, one like that Antoniadi experienced on September 20, 1909, when he thought he was dreaming and saw Mare Tyrrhenum "like a leopard skin." Rev. T. E. R. Phillips, previously a strong advocate of the use of small telescopes for planetary work, came to Meudon in 1911 and had the chance to look at Jupiter with the great telescope (see Fig. 15.3). It was only a fair night,

> definition was not perfect . . . but, though somewhat unsteady, the image was beautifully sharp at odd moments. The view of the planet in the great telescope was quite a revelation and a sight not to be forgotten. . . . Spots which I had previously observed as *simple* were shown to possess a highly *complex* structure, and the belts appeared more irregular and knotted than I had ever previously seen them. It was abundantly clear that there is a vast amount of minute detail on the planet which can only be revealed by instruments of great separating power.
>
> To sum up, the view of Jupiter was undoubtedly the finest I have ever had, and I can no longer question the immense value of great telescopes in the study of the planets. A large aperture shows them as a small aperture cannot possibly do, and any seeming uniformity or regularity which the latter may appear to indicate is at once dispelled. . . . The equatorial zone of Jupiter, e.g., was

*Figure 15.3
Drawings of Jupiter by E. M.
Antoniadi with the 32 3/4-inch
Meudon refractor: May 22, 1911
(above), and June 12, 1911 (below).*

seen to be a perfect maze of mottlings and soft wispy markings, but a striking feature of these markings was their irregularity.[69]

Impressive as such testimony is, I doubt that Lowell would have changed his mind even had he spent a whole season observing Mars at Meudon. Lowell's furrow was by this time too deep. Though he himself said, at the conclusion of his speech to the British Astronomical Association, "I am always skeptical myself, because I am not going to be found wrong," the word *skeptical* is perhaps not the first that would occur to someone wishing to describe him. Yet one can certainly accept the determination of the last part of his statement at face value: "I am not going to be found wrong." Though Antoniadi and Phillips had both come to acknowledge the mistakes they had made on the basis of observations with smaller telescopes, Lowell did not share their mental plasticity. Though he seems to have had the capacity for reaching decisions at lightning speed, once his mind was made up, it was made up. At the end of his career his stated views were in almost every respect indentical to those with which he had begun. He did not easily admit mistakes, and even less would he concede defeat. It is somewhat ironic, then, that in one of his last addresses he said:

> Youth is the period of possibilities. Then it is that the mind is open, plastic to impressions which at the same time it is most potent to retain. . . .
> Plasticity of mind is the premise to possibility of performance. To retain it longest is essential to success. . . . We are told that the good die young; our regret being father to the thought. But certain it is that the great die young even though they pass the Psalmist's limit of three score and ten. The plasticity of their makeup is the elixir of life poor Ponce de Leon sought in vain.[70]

By this standard, Lowell cannot be called great. On the other hand, in the same address he quoted the words of a Harvard commencement speaker: "Fellows, don't be content to sit on the fence; sit on the roof. And remember that climbing there does not safely consist in leaps and bounds but in throwing one's heart upward and the persistently pursuing it step by step."[71] So Lowell had done.

Schiaparelli, like Galileo blind in his old age, suffered a stroke in June 1910, from which he never recovered; he died on July 4. In his obituary of the great Italian observer, Lowell touched on a theme that was to become more and more pervasive in his own writings of later years—the persecution that the true pioneer had always experienced. Yet his words described his own situation, no doubt, more than they described Schiaparelli's: "Unwittingly the world pays to its pioneer minds the tribute of distrust. To nescience the advances of science are distasteful in proportion as they are correct. . . . No one likes his beliefs overthrown nor what he thought he knew upset, and the louder the outcry the more symptomatic is it of defeat."[72] As so often in Lowell's career, what he said redounded on his own head as much as it did on those of his critics.

As Mars retreated through a cycle of less favorable oppositions, finally bottoming out with that of 1916—Lowell's last— he turned his attention to other astronomical projects. His search for a trans-Neptunian planet was another disappointment to him.[73] But in his observations of the rings of Saturn, in which he made out a series of fine divisions (Fig. 15.4), the agreement he obtained between theory and observation could hardly have been more precise, a fact he must have found gratifying.[74] Unfortunately, it must be admitted that whatever the status of the markings that he saw in the rings, the fact that they showed up in the expected positions must be regarded as but another instance of the power of suggestion to influence such observations.

His comments about Mars in the year before his death on November 12, 1916, of an intracerebral hemorrhage, are worth recording. To Waldemar Kaempffert of the *Scientific American* he wrote: "The chief trouble with Antoniadi is that he is a man without knowledge of how to observe."[75] And in some of the last remarks he wrote about Mars, he confidently summed up his position:

> I have said enough to show how our knowledge of Mars steadily progresses. Each opposition as it comes round adds something to what we knew before. It adds without

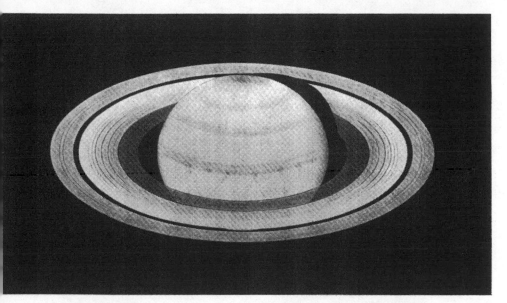

*Figure 15.4*
*Saturn and its rings, 1915,*
*showing the numerous fine*
*divisions in the rings Percival*
*Lowell and E. C. Slipher observed*
*with the 24-inch refractor. Lowell*
*found excellent quantitative*
*agreement between the positions of*
*his divisions and those he predicted*
*on the basis of the "resonance*
*theory," proposed some years earlier*
*by Daniel Kirkwood to explain*
*Cassini's division. The theory*
*held that areas of the rings where*
*particles moved in orbits with*
*periods that were integral ratios*
*of those of satellites—especially*
*Mimas, the innermost—would be*
*cleared out by the periodic tidal*
*disturbances of these satellites*
*and leave thinned-out zones or*
*actual gaps.*

subtracting. For since the theory of intelligent life on the planet was first enunciated 21 years ago, every new fact discovered has been found to be accordant with it. Not a single thing has been detected which it does not explain. This is really a remarkable record for a theory. It has, of course, met the fate of any new idea, which has both the fortune and the misfortune to be ahead of the times and has risen above it. New facts have but buttressed the old, while every year adds to the number of those who have seen the evidence for themselves.[76]

Actually, Lowell may have summed up his career better in these words, written long before in *The Soul of the Far East*: "Only the superficial never changes its expression; the appearance of the solid varies with the standpoint of the observer. In dreamland alone does everything seem plain, and there all is unsubstantial."[77]

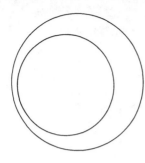

## CHAPTER 16

# Planets and Perception

*A glimpsed object is not as certain as an
object held steadily, and, however self-
evident or trite such a remark may be,
yet it is a very important one to make
here. . . . Were due regard to have
been paid to the treacherous character
of glimpsing, the existence of many
celestial marvels would never have
been foisted on the scientific world.*
                        —E. M. Antoniadi

IT WOULD BE unfair not to men-
tion that Lowell accomplished a great deal in astronomy in
addition to his work on Mars. An able celestial mechanician,
he carried out calculations that would oneday lead (albeit ser-
endipitously) to the discovery of the planet Pluto. His theories
of planetology, or the evolution of the planets, though premature, anticipated a branch of science that is only now ripening
in the age of spacecraft and that had one further, unexpected
result. While following up on Lowell's hunch that the "white
nebulae" were solar systems in formation (along the lines
of Laplace's nebular hypothesis), his assistant, V. M. Slipher,
made one of the premier discoveries of the twentieth century:
his spectroscope first revealed the large red-shifts of these
objects (now called "galaxies"), the crucial finding leading to
the recognition of the expansion of the universe. Finally, the
Lowell Observatory, directed by Slipher after Lowell's death,
has proved an enduring legacy to the energy and organizing
genius of the man. Yet for better or worse, Lowell's name will
always be associated primarily with his visions of the canals of
Mars and the inhabitants of the planet he deduced from them.    *257*

WILLIAM JAMES, in his 1890 book *Principles of Psychology* (with which Lowell himself was familiar) wrote:

The object of belief, then, reality or real existence, is something quite different from all the other predicates which a subject may possess. Those are properties intellectually or sensibly intuited. When we add any one of them to the subject, we increase the intrinsic content of the latter, we enrich its picture in our mind. But adding reality does not enrich the picture in any such inward way; it leaves it inwardly as it finds it, and only fixes and stamps it in to us. . . . The *fons et origo* of all reality, whether from the absolute or the practical point of view, is thus subjective, is ourselves. As bare logical thinkers, without emotional reaction, we give reality to whatever objects we think of, for they are really phenomena, or objects of our passing thought, if nothing more. But, *as thinkers with emotional reaction, we give what seems to us a still higher degree of reality to whatever things we select and emphasize and turn to* WITH A WILL. These are our *living* realities; and not only these, but all other things which are intimately connected with these. Reality, starting from our Ego, thus sheds itself from point to point—first, upon all objects which have an immediate sting of interest for our Ego in them, and next, upon the objects most continuously related with these. It only fails when the connecting thread is lost. A whole system may be real, if it only hang to our Ego by one immediately *stinging* term.[1]

What James says here applies, of course, to the features on planetary surfaces as to anything else. For Lowell the existence of life on Mars certainly hung to the Ego by immediately stinging terms. In the end, of course, he was proved wrong, but the passionate advocacy of his theory was to cause more than a few young minds to turn with a will to astronomy and to lead, not very indirectly, to Mariner and Viking. Paradoxically, the soberer and lower-flying empiricists, who might be regarded by some as better role models by whom to frame the scientific character, have often left others cold to their subject. Beer and Mädler come at once to mind. As Agnes M. Clerke wrote of their great work *Der Mond,*

This summation of knowledge in that branch, though in truth leaving many questions open, had an air of finality which tended to discourage further inquiry. It gave form to a reaction against the sanguine views entertained by Hevelius, Schröter, Herschel, and Gruithuisen as to the possibilities of agreeable residence on the moon, and relegated the "Selenites," one of whose cities Schröter thought he had discovered, and of whose festal processions Gruithuisen had not despaired of becoming a spectator, to the shadowy land entered through the Ivory Gate. All examples of change in lunar formations were, moreover, dismissed as illusory. The light contained in the work was, in short, a "dry light," not stimulating to the imagination. "A mixture of a lie," Bacon shrewdly remarks, "doth ever add pleasure."[2]

The quote from Bacon is from his essay "Of Truth." There he adds: "Doth any man doubt that if there were taken out of men's minds vain opinions, flattering hopes, false valuations, imaginations . . . and the like, but it would leave the minds of a number of men poor shrunken things, full of melancholy and indisposition, and unpleasing to themselves?"[3]

The public imagination will always be more captivated by the sensational aspects of science than by the routine work which, some might argue, is of far greater importance. Comets, celestial lightweights, swoop in once in a great while, stir up the public, grab all the headlines, and then vanish back into empty space. The public has more interest in Lowell's Martians (or in any of their later permutations) than in many a more modest branch of celestial science. All that glitters is not gold: the same public is notoriously gullible, and science must share the stage with astrology and the occult. Yet a comet may for a moment inspire a sense of wonder and of belonging to a larger whole, and Lowell's Mars, if a fairy tale, at least "aroused generations of eight-year-olds," as one so aroused has pointed out.[4] Of them it is still true, as Samuel Johnson once said, that "they like to be told about giants and castles, and of somewhat which can stretch and stimulate their . . . minds."[5]

We now know that there are no canals dug by Martians, and no Martians either. The spacecraft photographs have proved that. Yet there is one last question: What, then, did Lowell spend twenty-two years watching and drawing?

IN LOWELL'S DAY, perceptual psychology was yet in its infancy as a scientific discipline. Lowell himself made a minor contribution to the subject through his 1903 experiments on the threshold visibility of telegraph wires. He and his assistants, V. M. Slipher and C. O. Lampland, found that such wires could be seen with the naked eye when their angular diameter was only 0.″69 of arc.[6] This meant that on Mars as seen through the 24-inch refractor, the eye ought to be able to detect linear features only 3/16 of a mile across. The idea that irrigation channels, or at least the strips of vegetation growing alongside them, would be visible with optical instruments from the Earth was, therefore, in Lowell's view, completely tenable.

This gives some idea of what Lowell thought he was seeing. He believed he could make out the surface structures of the planet—at least with respect to linear forms—to a resolution of less than a mile. Lowell himself speculated on how the eye was able to achieve its truly extraordinary discrimination of lines. It was due, he said, to "summation of sensations." "What would be far too minute an effect upon any one retinal rod [i.e., cone] to produce an impression," he wrote, "becomes quite recognizable in consciousness when many in a row are similarly excited."[7]

Lowell had started out with the conviction that the lines he saw on Mars were genuine surface features; only later did he seize on the eye's capacity for detecting fine lines to bolster this belief. E. Walter Maunder had come to the same realization about the eye's power as a line detector long before Lowell but had arrived at it by an entirely different route. As mentioned earlier, during the course of his solar observations he discovered that "a straggling group of small unimportant spots" could be made out even though much smaller in actual area than a single spot. At the same time, a simple calculation showed that even an apparent point on the surface of the Sun was still a widely extended area and might even be larger than the Earth. Thus his conclusion: "We have no right to assume, and yet we do habitually assume, that our telescopes reveal to us the ultimate structure" of a planet.[8]

The difference, then, was that whereas the canalists seem to have had a rather literal faith in what was revealed to them by the senses, Maunder recognized that what appeared to be was not necessarily what was. Far from it. The telescope no more revealed the ultimate structure of a planet than the naked eye did the ultimate structure of, say, an infusorian, which though appearing as a simple dot is found under magnification to be made up of parts, each of which may under still higher magnification show still more delicate structure. Schiaparelli, pondering the regularity of various natural phenomena, waxed eloquent on the beautiful geometry of nature:

> The geometry of nature is manifested in many facts . . . from which are excluded the idea of any artificial labor whatever. The perfect spheroids of the heavenly bodies and the ring of Saturn were not constructed in a turning lathe, and not with compasses has Iris described within the clouds her beautiful and regular arch. And what shall we say of the infinite variety of those exquisite and regular polyhedrons in which the world of crystals is so rich! In the organic world, also, is not that geometry most wonderful which presides over the distribution of the foliage upon certain plants, which orders the nearly symmetrical, starlike figures of the flowers of the field, as well as of the animals of the sea, and which produces in the shell such an exquisite conical spiral, that excels the most beautiful masterpieces of gothic architecture? In all these objects the geometrical form is the simple and necessary consequence of the principles and laws which govern the physical and physiological world. That these principles and these laws are but an indication of a higher intelligent power, we may admit, but this has nothing to do with the present argument.[9]

Much in nature does indeed appear thus regular. Yet much that appears so does so only because we are looking at it in the broad view. Examined closely enough, every regularity betrays its underlying irregularity.

WE NOT ONLY visually gloss over irregularities, however; we tend actively to impose order and regularity where it does not exist—a recognition that Maunder, Cerulli, and others were groping toward in their attempts to understand the phenomena of Mars and that later became fundamental to the Gestalt school of perceptual psychology, of which they may be regarded as forgotten predecessors. A perception is not born, it is made. The Gestaltists were later to stress the importance of the perceptual tendency to optimize "figural goodness." Whereas under favorable conditions—good illumination, long exposure, direct foveal stimulation, and the like—the perception tends to be determined primarily by external forces, when these are lacking, internal forces play a more significant role— for instance, tendencies toward continuity and figure completion. Consider the image shown in Figure 16.1. Though the triangle is incomplete, one sees definite, continuous contours even where one knows that none actually exist.[10] These have been called, fittingly enough, subjective contours, yet though one knows that the lines are not really there, the recognition itself does not serve to dispel the impression of them. One knows that they are illusions, yet one nevertheless sees them. No further elaboration of the point is necessary, but certainly the propensity of the dark areas of Mars to form pointed or caret-shaped bays—the sources of many of the canals—suggests that subjective contours may have been responsible here also, at least for some of the more delicate components of the Martian spiderwebs.

The tendency for internal forces to shape perception is evident in the history of Martian observation long before the canals enter the picture, however. One already sees it, for instance, at the stage where the planet's surface showed nothing more than coarse differentiation into figure and ground. In ambiguous figures such as the Necker cube (a box drawn in skeletal outline so that one can see it in either of two ways) the stimuli are so arranged as to evoke with equal or nearly equal probability two different responses. The balance between these is a precarious one, and the mind shifts readily from one to the other. It is best to reserve the term *ambiguous figure* for cases such as these, where the alternative presentations are thus balanced. In other cases the figures are better described as unstructured or relatively meaningless. Lacking any well-defined meaning in themselves, the mind has relatively free play to project its own meanings onto them. Rorschach, or inkblot, patterns are of this type, and the psychological tests in

**Figure 16.1**
*Kanizsa's figure illustrating
"subjective contours." The brain
receives—or creates—the
impression of faint lines to complete
the figure, even though the areas in
which the lines appear are in fact
blank.*

which they are used are accordingly called projective tests. In practice, the interpretations of a given Rorschach pattern tend to cluster around the few most obvious possibilities, and these tend to be regarded as more or less the "normal" way of seeing them, as opposed to the highly idiosyncratic way that psychotics or very imaginative people may see them. In the interpretation of the splotches on the Moon and Mars, it is not surprising that something similar should have been true; by habit and past association, they tended most often to be seen in terms of the (projected) analogy to terrestrial features—oceans and lands.

Once astronomers got beyond that stage of the telescopic observation of Mars that was confined to the mere recording of these "splotches," still more interesting psychological effects came into play. The small details on what were believed to be the continental areas of Mars were first recognized in profusion by the keen-sighted Schiaparelli in 1877. Yet with his 8.6-inch refractor these features were just at the threshold of detection. One needed a perfectly steady atmosphere to discern them at all, yet generally the atmosphere cooperated only in brief flashes.

Arthur A. Hoag has written the following concise summary of the atmospheric factors involved in setting the limits to telescopic performance on a given night:

> The resolution in an image is related to the nature of the wavefront at the telescope aperture in the case of a perfect telescope. Diffraction theory then dictates the resolving power, which is inversely proportional to the aperture. In the practical case, the resolution is set by the instantaneous size of coherent patches over the aperture. At a good site these patches average 10 to 20 cm in size. The distribution in sizes is such that coherence over the whole aperture of a 24-inch telescope is rare, which is why Lowell and others would stop down their apertures. However, the full resolving power can sometimes be realized for short instants. . . . For this reason there is an increasing tendency to characterize seeing by *Fried's parameter,* $r_o$, which is the average size of wavefront disturbances. For the typical $r_o = 12$ cm the resolving power is about 1 arcsec. If you wait long enough, a flash will come along

The periods during which the wavefront becomes coherent over the whole aperture typically last on the order of a fraction of a second. Thus seeing is by glimpses, or as Percival Lowell once put it, by "revelation peeps." The effect is like that produced by a tachistoscope, a device used by perceptual psychologists since the turn of the century to study what takes place during brief perceptions. By analogy, I shall refer to the atmosphere's usual tendency to allow excellent views in only brief snatches as the tachistoscope effect.

Again and again, as one reads through the accounts of visual observers telling how they came to see the canals, one finds a consistent theme. Percival Lowell wrote in his notebook concerning the first canal he saw in 1894: "In the glimpses in which it was caught sometimes thought it more inclined to the north." Later he thought he saw it double but under much the same circumstances: "Glimpsed double. These sudden revelation peeps may or may not be the truth."[12]

Clearly, planetary seeing involves a process of trying to make out "infinite riches in a little room," to borrow a phrase from poet Christopher Marlowe—the disk, after all, is exceedingly small—but just as important, of trying to make out these infinite riches in a little time. Lowell admitted in his book *Mars* that "the steadiest air we can find is in a state of almost constant fluctuation. In consequence revelations of detail come only to those who patiently watch for the few good moments among the many poor. . . . To see [the canals] as they are . . . an atmosphere possessing moments of really distinct vision is imperative."[13] It is with glimpses and "revelation peeps" that we have to do here. Expectation, needless to say, is sure to play a fairly large role. Imagination fleshes out the fleeting impression. "Suspected . . . multitudinous canals all too fugitive to be positively figured," Lowell entered in his notebook one night. The next night, this became "network of canals . . . general effects quite like Schiaparelli's globe."[14] The transformation from one to the other is the very process with which we are here most concerned. For a start, we may recall the insightful comment by Rev. T. E. R. Phillips, based on his own experience in observing Mars a few years later:

"Faint markings have been glimpsed now and then, and it would be easy, by the 'scientific use of the imagination,' to conjure them into lines and streaks harmonising with Schiaparelli's charts." But being more cautious than Lowell, he added: "I am careful to represent on my drawings only what I feel I can see with certainty and hold with tolerable steadiness."[15]

One waits for the tachistoscope flash, and one waits with a definite expectation in mind of what might be revealed in it. Indeed, one must. Planetary markings, except for the coarsest, do not usually hit one over the head with their obviousness, yet it is important to keep the mind open. If one commits oneself too early to any one possible interpretation, as Sir Ernst Gombrich says concerning the interpretation of "whiffs" of sound coming over the radio, it may be impossible henceforth to see it any other way.[16] The analogy between visual planetary observation and Gombrich's monitoring of weak and static-filled radio transmissions is not as farfetched as may at first appear. Both cases involve an observer trying to make sense out of a noisy and degraded transmission; the planetary observer is concerned with a more or less blurred visual message, the radio monitor with a fragmentary auditory one. The sensory modalities are, however, the only thing that is different; otherwise, the process is exactly the same.

Perhaps no one has given a better description of the anatomy of visual planetary observation than Percival Lowell himself when he wrote in his 1906 book *Mars and Its Canals*:

> When a fairly acute-eyed observer sets himself to scan the telescopic disk of the planet in steady air, he will, after noting the dazzling contour of the white polar cap and the sharp outlines of the blue-green seas, of a sudden be made aware of a vision as of thread stretched somewhere from the blue-green across the orange areas of the disk. Gone as quickly as it came, he will instinctively doubt his own eyesight, and credit to illusion what can so unaccountably disappear. Gaze as hard as he will, no power of his can recall it, when, with the same startling abruptness, the thing stands before his eyes again. Convinced, after three or four such showings, that the vision is real, he will still be left wondering what and where it was. For so

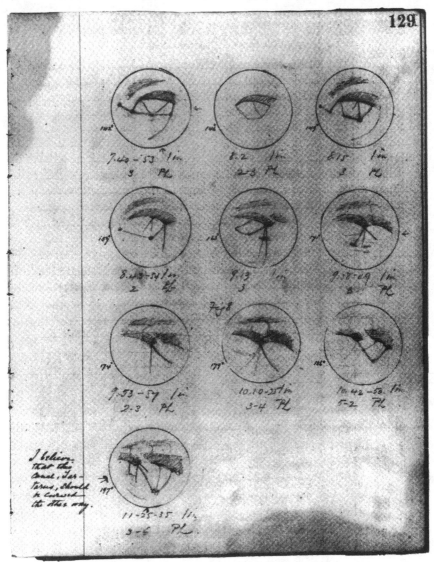

*Figure 16.2*
*Impressions of Mars by Percival Lowell. Representative page from his observing logbook, 1894, showing how the canals appeared to him during the momentary flashes of good seeing. This illustrates the concept referred to in the text as the tachistoscope effect.*

short and sudden are its appearances that the locating of it is dubiously hard. It is gone each time before one has got its bearings.

By persistent watch, however, for the best instants of definition, backed by the knowledge of what he is to see, he will find its comings more frequent, more certain and more detailed. At last some particularly propitious moment will disclose its relation to well known points and its position be assured. First one such thread and then another will make its presence evident; and then he will note that each always appears in place. Repetition *in situ* will convince him that these strange visitants are as real as the main markings, and are as permanent as they.[17]

Others saw the canals in much the same way. E. M. Antoniadi, writing of his observations in 1894 with the 9½-inch refractor at Juvisy, reported: "I everywhere saw single or double lines for an eighth of a second or so."[18] On another occasion he wrote that he was "innured for years to the fleeting visibility of straight lines on Mars with the ordinary appliances" he was using.[19] Massachusetts engineer Elihu Thompson, in a letter to Lowell's friend E. S. Morse, said that he managed to see some of the canals in a 10-inch reflector, "and *at times, only at times,* a network appearance of them. . . . The proof that this is no illusion consists in the fact that only at the moments of the very greatest steadiness could the detail be seen:—canals and all. If the effect were the result of an optical illusion, it should have appeared . . . even when the disc was only fairly steady."[20] The South African observer Lindsay A. Eddie recorded his impressions of Mars in 1907: "Many reticulations and differential shadings were occasionally glimpsed in fleeting moments of good seeing, too brief, alas! to be fixed or reproduced."[21]

Among more recent observers, Robert J. Trumpler, observing Mars in 1924 with the great Lick refractor, reported that certain areas on the planet "made, most of the time, the impression of a slightly shaded area. But when, for a moment, the definition improved they suddenly resolved into a maze of well-defined lines, too complicated, however, to be retained by the memory from such short flashes, even if they repeated themselves."[22] Clyde Tombaugh, who often saw canals with

Lowell's own telescope, found that the canals "came out in brief flashes (less than a second of time) whenever larger air-cells of uniform refractive index pass in front of the telescope aperture."[23] And so on ad nauseam.

Nathaniel Green thought that the canals sometimes corresponded to the boundaries of regions of different shade or tone, at other times to the eye's tendency to draw out certain baylike features into wispy lines. The "subjective contour" interpretation of the canals was later revived by W. H. Pickering, who wrote in 1904: "The chief cause of the illusion seems to be the system of lakes, or oases as they are sometimes called, which were first discovered in large numbers at Arequipa. There is a curious tendency of the human eye to see such dark points united by faint narrow lines, and it has been shown by means of diagrams that these lines sometimes appear when the diagram is at such a distance that the dark spots are themselves invisible."[24] Maunder accepted both of these explanations and added to them the eye's tendency to join together minute dotlike markings, though in fact, as he pointed out, "it is not necessary that the markings should be dot-like. . . . They may be of any conceivable form provided only that they are outside the limit of distinct vision and are sufficiently sparsely scattered."[25]

Clearly, the eye is quite capable of producing the impression of illusory lines by any of a number of mechanisms, its propensity for doing so being no doubt related to its phenomenally low threshold for seeing *real* lines. Thus the gist of Maunder's rebuttal of Lowell's deduction from telegraph-wire experiments that he was seeing lines on Mars of the order of a mile or so in breadth was that

> since there was so wide a difference between the limit at which we have real definition of the object and the limit of mere perception, even though it be distinct, it followed that markings well within those limits . . . must tend to assume the simplest geometrical form—the forms, that is to say, of the circular dot and the straight line.[26]

Maunder's point is that the eye does not produce a literal transcription of reality. This is particularly true under difficult conditions of observation. The image is schematized. The eye enhances the borders of contrasts; it sharpens boundaries blurred by diffraction and chromatic aberration.[27] The lens of the eye is far from homogeneous and is optically quite imperfect. Yet the retina largely compensates for the inadequacies of the lens. It presents a reconstructed image. But when the information to be extracted does not lie far above the threshold of "noise," it can become difficult to sort out an artifact due to visual processing from the features of the object being observed. Schiaparelli was accused of turning "soft and indefinite pieces of shading into clear, sharp lines." His drawings appeared to show Mars in "plan" view; the Martian markings seemed to be presented under the forms of an "unpleasantly conventional" style.[28] Yet everything that Schiaparelli's critics said about how he reported what he saw on Mars could equally well be imputed to what his eye reported to him, for the eye can do all of this. It is not interested in how things "really" look, but only in seizing on those aspects of a stimulus which may be of most interest to it. It so happens that geometric forms seem to play a dominant role here. The visual system has hardwired into it components that act as preferential filters to pull certain specific features out of the stimulus.[29] There is a sense, then, in which the ideal forms about which Plato spoke so long ago *do* exist, not as entities subsisting in some pure Platonic heaven, but hardwired into the visual system and making up what might be referred to as the "blanks and formularies" of perception.[30]

In the short tachistoscope flashes at the telescope, the eye catches, now and again, something fleetingly present on the planetary disk, and it attempts to give to this "a local habitation and a name." The novice is apt to feel bewildered as to the precise shape or position of what was vouchsafed in so brief a revelation. With more experience, however, the eye's agility improves and the glimpses become more assured. The experienced observer can be said to have become a speed reader of planetary details, able to build up his or her interpretation in terms of whole phrases, or what passes for such on the planetary disk. Schiaparelli may have been born keensighted, but only by training did he become swift sighted, and the way he did so was the way that every visual observer of the planet must do so. The eye must commit what is sees to shorthand.

Perceptual psychologists have known for a long time what takes place when the eye is allowed only tachistoscopic glimpses in which to make out a stimulus. Of particular interest here is the experimental work of John H. Flavell and Juris Draguns, performed during the 1950s on what they termed the microgenesis of perception: the sequence of events assumed by them to take place between the presentation of a stimulus and the formation of a stable percept. Normally these "precursors" are not observed; by selectively blocking complete perception, however, as by presenting visual stimuli tachistoscopically with sequentially increasing exposure times, Flavell and Draguns believed that they could in fact be elicited:

> The initial perception is that of a diffuse, undifferentiated whole. In the next stage figure and ground achieve some measure of differentiation, although the inner contents of the stimulus remain vague and amorphous. Then comes a phase in which contour and inner content achieve some distinctness and a tentative, labile configuration results. Finally, the process of Gestalt formation becomes complete with the addition of elaborations and modifications of the "skeletal Gestalt" achieved in the previous stage.
>
> As development proceeds, external, objective characteristics more and more supplant inner, personal factors as determinants of the structure perceived. . . . Of particular interest . . . [is] the stage just preceding the formation of the final, stable percept. In this *Vorgestalt* or preconfiguration phase the subject has constructed a tentative, highly labile Gestalt which is . . . more regular, and more simple in form and content, than is the final form which is to follow it. The construction of this initial, fluxlike pre-Gestalt is said to be accompanied by decidedly unpleasant feelings of tension and unrest which later subside when a final, stable configuration is achieved. The emotionally charged character of the *Vorgestalt* stage is stressed by many investigators.[31]

The relevance of this to the Martian canals should be immediately obvious. Lowell, in an 1897 paper entitled

"Atmosphere: In Its Effect on Astronomical Research," described a sequence of stages of presentation of the Martian disk which is remarkably like that described by Flavell and Draguns:

> Taking the bad air with the good the planet presented a regular object lesson in atmospheric effect: showing itself at one time or another under the several aspects it has worn to previous observers in an ascending scale of definition tied to increasing steadiness of the air. I have thus seen the disk successively put on the look recorded by Huygens, Lockyer, Green, and Schiaparelli and so strikingly that I made note of it at the time.[32]

The stage in the "microgenetic unfolding" to which the canaliform Mars relates is evidently that referred to by Flavell and Draguns as the Vorgestalt, or "preconfiguration" phase, characterized by "a tentative, highly labile Gestalt which is . . . more regular, and more simple in form and content than is the final form which is to follow it." One notes, moreover, its emotionally charged character. Antoniadi spoke of the "fleeting apparition of *hideous* straight lines,"[33] while Schiaparelli in 1888 wrote:

> What strange confusion! What can all this mean? Evidently the planet has some fixed geographical details, similar to those of the Earth. . . . Comes a certain moment, all this disappears to be replaced by grotesque polygonations and geminations which, evidently, attach themselves to represent apparently the previous state, but it is a gross mask, and I say almost ridiculous.[34]

There *is* an objective basis for such fleeting impressions. The canalists were right in insisting that if the canals were a "mere" illusion, they should have been more indiscriminate about when they showed. They ought not to have picked only the moments of finest seeing to become visible in their small or stopped-down telescopes. Yet whereas the canalists took this as evidence that the lines themselves were real, in fact they should only have concluded that they had a *basis* in reality, for Lowell's series of stages corresponding to "increasing steadiness of the air" ought not to have stopped with Schiaparelli. To refer again to Arthur Hoag's comments about atmospheric

seeing, if one waits long enough and has a large enough telescope, eventually a tachistoscope flash will come along in
which the wave front is coherent over a diameter many times
larger than average. It is in these moments that the superiority
of the large telescope asserts itself, and the true basis of the
apparent lines becomes visible. Antoniadi wrote of his experience with the 32³/₄-inch: "Its superiority on a good night is
beyond question; under bad seeing . . . its power [will still]
make itself felt during the fleeting moments of comparative
calm." Phillips wrote of his views of Jupiter with the same
telescope on a night that was only fair that "though somewhat
unsteady, the image was beautifully sharp at odd moments"[35]—the tachistoscope effect again, but this time with
enough resolving power to make out the details more clearly.

A number of investigators have concluded that, with few
exceptions, the canals were without objective basis, pointing to
the poor correlation of the charted canals with aligned details
in the spacecraft photographs.[36] The claim is based on attempts to overlay maps of canals on spacecraft maps. But their
results are not conclusive. Some of the spacecraft maps used
do not seem to have been quite satisfactory for the purpose;
while showing relief features, they did not always show albedo
features adequately, yet it is among the latter that we must
look for the most likely objective stratum of the canals. In any
case, these are changeable over time. The best Earth-based
photographs do, however, show unmistakably a delicate mottling of the surface, a kind of roughly interlacing tracery of
details, which corresponds to the descriptions that many
skilled observers have given over the years. What this amounts
to, in fact, is a blurred view of the multitude of streaks,
splotches, spots, and craters shown in the spacecraft photographs and maps (Fig. 16.3). Except in a few cases, the canals
do not, it is true, correspond to these underlying features in
any straightforward way; they are not, in most cases, a mere
"summing up" of roughly aligned features too small to be
individually distinguished (as Maunder too simplistically proposed). On such a disk, viewed under such difficult conditions
of observation, we have no right to expect the eye to be even
as literal as that. But the complex features are there, and there
can be little doubt that the canals represent an impression—
and hardly a precise one—of these features in the form of the

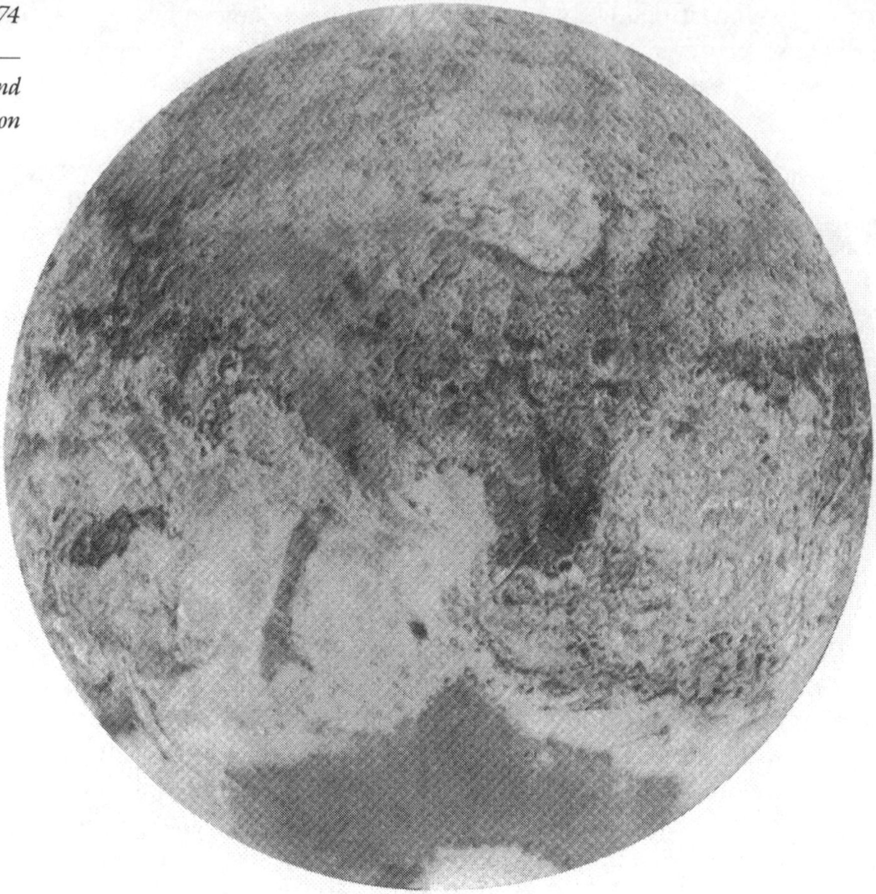

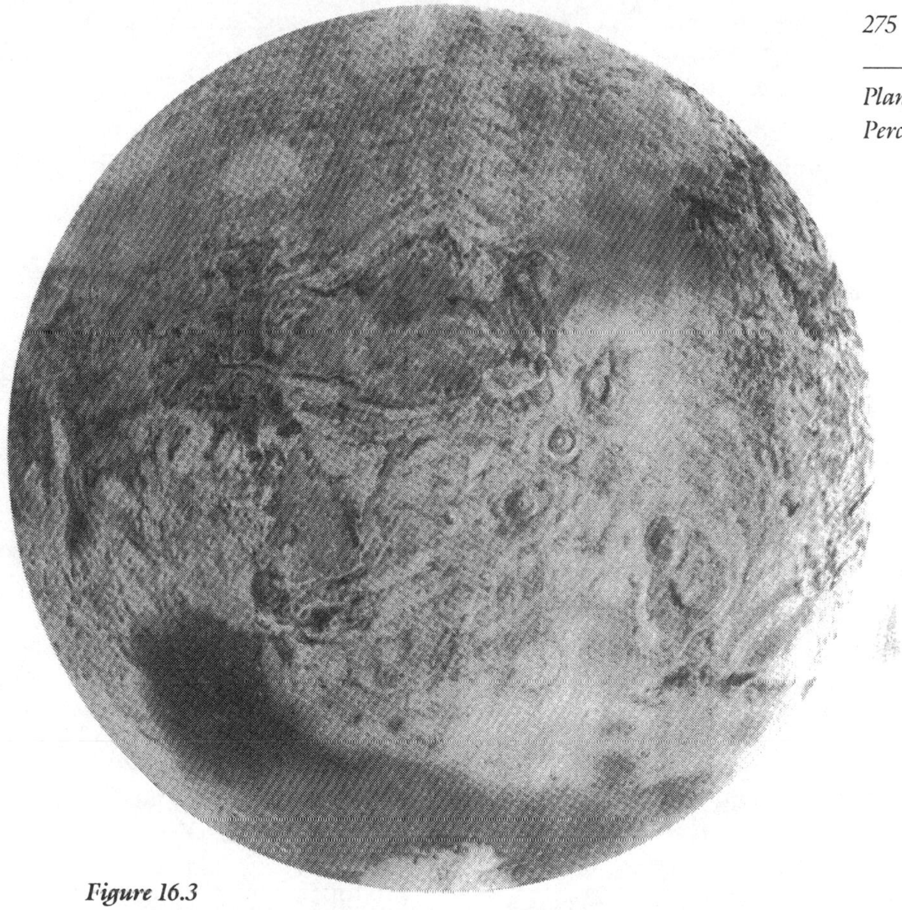

**Figure 16.3**
*Modern charts of Mars, showing
complementary hemispheres and
based on spacecraft photographs.
The charts are inverted, with south
at the top, to facilitate comparison
with telescopic views of the planet.*

*The chart on the left is centered
nearly on the conspicuous dark
area Syrtis Major, and that on the
right shows the gigantic canyon
Valles Marineris and the volcanoes
of the Tharsis region.*

eye's own built-in shorthand of lines, circular spots, and geometrical forms generally.

Antoniadi realized, on the basis of his views with the Meudon refractor, that the geometrical canal network was an optical "symbol" and that in its place there were "myriads of marbled and chequered objective fields which no artist could ever think of drawing."[37] Wherever Schiaparelli had drawn canals, he found, "as far as it was possible to judge . . . a group of irregular shadings on the surface of Mars."[38] Thus he concluded that

> the student who passes many consecutive hours in the study of Mars with medium-sized instruments, is liable to catch rare glimpses of straight lines, single or double, generally lasting about a quarter of a second. Here we have a vindication of Schiaparelli's discoveries. But their deceitful character will obtrude itself on the observer using a large telescope when, in place of the lines, he will hold steadily, either a winding, knotted, irregular band, or the jagged edge of a half-tone, or some other complex detail.[39]

There is little more to say. There is no question that Lowell was right about the artificiality of the canals of Mars. The canals were, however, artificial not because the Martians made them but because what we perceive is itself in an important sense an artifact. Schiaparelli and Lowell took the Vorgestalt of the perception—essentially, a mere schema invented by the eye in order to make the image rather artificially hang together—for the planet's ultimate structure. Perceptually speaking, they were fundamentalists; they took what the eye reported to be the literal truth. Yet, alas, there was more to Mars than met the eye—or at least, more to it than met their eyes. Theirs was in many ways a simpler world. They still believed in a neat order and a mathematical regularity to the universe, perhaps partly as a result of the illusion that makes things at a distance seem, in Schiaparelli's words "as if made by rule or compass" or "rounded on a lathe."

The Meudon telescope in Antoniadi's hands, like the Lick telescope in Barnard's a decade and a half before, showed clearly that underlying the apparent simplicity of the surface of Mars was a complexity of such an order that it was simply, to use Antoniadi's word, bewildering. Two years later Phillips saw similar complexity in the belts of Jupiter. Since then we have seen those planets still more clearly in the spacecraft photographs, discovering still more minute and intricate complexity enfolded within the complexity that they saw. The planets become Chinese puzzles of complexity. Having only begun to absorb Antoniadi's drawings, once again we find ourselves bewildered. In the same way, the rings of Saturn, which Lowell and others of his day regarded as a striking testament of a relatively simple and changeless celestial order, have for us, with the *Voyager* spacecraft, been broken up into a series of innumerable intricate ringlets that reflect complex processes and that are probably subject to considerable change, perhaps even over fairly short intervals of time. Rather than emblems of eternity, planetary rings appear to be relative transients, celestially speaking, their ages reckoned in millions of years, compared to billions of years in the case of the planets themselves.

Percival Lowell once wrote in his observing notebook: "The best night I have had. On mounting stairway to clock found the sky an exquisite pale liquid blue, with Mars orange upon it." Thus the planet looked to the ancients, a mere burning coal upon the night sky. In Huygens's telescope, it first began to resolve into apparent continents and seas, becoming thereby a distant Earth, yet one invested with a certain aura of ideality, perhaps, like all pastures seen at a distance, which are apt to seem greener and more lush than our own. Schiaparelli, Antoniadi, and the spacecraft photographs each in turn have introduced a new level of complexity into our view of it so that now we have entered on the fifth or still higher level in our perception of the complexity of Mars and, in the process, have seen it divested more and more of the quality of dream and invested more and more with that of solid reality. It is a place now, with sunrises and sunsets, and solid, kickable rocks. The next phase in our recognition of this new world will come when we actually take our place there as inhabitants, temporary or permanent, when we ourselves become Martians. No doubt the world that we find there—and that we create there—in time will become every bit as complex and befuddled as that which we will be escaping from. And a persistent human tendency will no doubt follow us there as well—our remarkable penchant for oversimplification. Lowell might not have found the planet as it appears to us in the spacecraft photographs to be more exciting than the world he created in his extraordinarily vivid imagination. But Maunder would have enjoyed it. It was he, after all, who had said all along: "We have no right to assume, and yet we do habitually assume, that our telescopes reveal to us the ultimate structure of the planet."[40]

# NOTES

*Chapter 1.*
*Victorians, Apes, and Martians*

1. The inscription is taken from Lowell's *Mars and Its Canals* (New York, 1906), p. 7; ellipsis in original.

2. Alfred, Lord Tennyson, *In Memoriam*, sec. xxi, lines 18–20.

3. Paley is quoted in Sir Gavin de Beer, "Biology before the *Beagle*," in *Darwin*, ed. Phillip Appleman (New York, 1970), p. 9. Percival Lowell, *Mars as the Abode of Life* (New York, 1908), pp. 109–110.

4. Jean Piaget, *The Child's Conception of the World*, trans. Joan and Andrew Tomlinson (London, 1929), p. 14.

5. Charles Darwin, *The Descent of Man* (New York: Modern Library), pp. 919–920.

6. Tennyson, *In Memoriam*, sec. lvi, line 15.

7. Andrew Carnegie's remarks are fairly typical: "The price which society pays for the law of competition, like the price it pays for cheap comforts and luxuries, is also great; but the advantages of this law are also greater still than its cost. . . . But, whether the law be benign or not, we must say of it . . . it is here, we cannot evade it; no substitutes for it have been found; and while the law may sometimes be hard for the individual, it is best for the race, because it insures the survival of the fittest in every department." *The Gospel of Wealth, and Other Timely Essays*, ed. Edward C. Kirkland (Cambridge, Mass., 1962), p. 16.

8. Lowell, *Mars as the Abode of Life*, pp. 111–112.

9. Ibid., p. 135.

10. Quoted in William Graves Hoyt, *Lowell and Mars* (Tucson, 1976), p. 299.

11. This, for instance, was Freud's view; see *The Future of an Illusion*, trans. James Strachey (New York, 1961), p. 24: "When the growing individual finds that he is destined to remain a child for ever, that he can never do without protection against strange superior powers, he lends those powers the features belonging to the figure of his father."

12. Bernard de Fontenelle, *Entrietens sur la pluralité des mondes*, ed. Robert Shackleton (Oxford, 1955), p. 54.

13. Dale P. Cruikshank to William Sheehan, April 20, 1987.

14. Quoted in W. F. Denning, "Jupiter and the Relative Powers of Telescopes in Defining Planetary Markings," *Observatory*, 8 (1885), 79.

15. E. Walter Maunder, "The 'Eye of Mars,'" *Knowledge*, March 1, 1895, 57.

16. Sir Winston Churchill, *Painting as a Pastime* (London, 1948), pp. 28–29.

1. Plutarch, *De Facie Quae in Orbe Lunae Apparet (περὶ τοῦ ἐμ-φαινομείον προβώπον τῳ κύκλῳ τῆς θελείης)*, 920F. The translation used here is that of A. O. Pickard (London, 1911).

2. *Plutarch's Moralia,* trans. Harold Cherniss and William C. Helmbold, Loeb Classical Library (Cambridge, Mass.: 1957), vol. 12, p. 48n.

3. Plutarch, *De Facie,* 922A.

4. Johannes Kepler, *Kepler's Conversation with Galileo's Sidereal Messenger,* trans. Edward Rosen (New York, 1965), p. 13.

5. Alexander von Humboldt, *Kosmos* (Stuttgart, 1850), vol. 3, p. 544.

6. Concerning Gilbert's drawing, see Ewen A. Whitaker, "*De Facie in Orbe Lunae,*" *Astronomical Quarterly,* 1 (1977), 61–65.

7. Galileo Galilei, "The Starry Messenger," in *Discoveries and Opinions of Galileo,* trans. Stillman Drake (New York, 1957), p. 32.

8. Ibid., pp. 33–34.

9. Kepler, *Conversation,* pp. 26–27.

10. For Galileo's influence on Harriot, see Terrie F. Bloom, "Borrowed Perceptions: Harriot's Maps of the Moon," *Journal for the History of Astronomy,* 9 (1978), 117–122.

11. Quoted in Norwood Russell Hanson, *Patterns of Discovery* (Cambridge, 1958), p. 184.

12. Quoted in Bloom, "Borrowed Perceptions," pp. 119–20, who gives as her reference Henry Stevens, *Thomas Harriot: The Mathematician, the Philosopher, and the Scholar* (London, 1900), p. 116. Lower's orthography and punctuation have been modernized.

13. Bloom, "Borrowed Perceptions," p. 121.

14. Galileo, "Starry Messenger," p. 36.

15. E. H. Gombrich, *Art and Illusion: A Study in the Psychology of Pictorial Representation,* 2nd ed., (Princeton, 1961), p. 81.

16. Quoted in Giorgio Abetti, *Storia dell'Astronomia* (Florence, 1963), p. 106.

17. Quoted in A. F. O'D. Alexander, *The Planet Saturn: A History of Observation, Theory, and Discovery* (London, 1962), p. 86.

18. Christiaan Huygens, *Systema Saturnium,* in *A Source Book in Astronomy,* ed. Harlow Shapley and Helen E. Howarth (New York, 1929), p. 164.

19. R. L. Gregory, *The Intelligent Eye* (New York, 1970), p. 119.

20. Quoted in Rev. T. W. Webb, *Celestial Objects for Common Telescopes* (New York, 1962 reprint of 6th ed., 1917), vol. 1, p. 210.

*Chapter 3.*
*A Multitude of Earths*

1. Galileo to Giacomo Muti, February 28, 1616.

2. Galileo Galilei, *Dialogue Concerning the Two Chief World Systems,* trans. Stillman Drake (Berkeley, 1967), p. 99.

3. John Milton, *Paradise Lost*, bk. I, lines 288–291.

4. Kester Svendsen, *Milton and Science* (Cambridge, Mass., 1955), p. 73.

5. John Wilkins, *The Discovery of a World in the Moone; or, A Discourse tending to prove that 'tis probable there may be another habitable world in that planet* (London, 1638), pp. 95, 187.

6. According to Beer and Mädler, Riccioli's work would have been consigned to oblivion "had he not been led by vanity to find a place on the Moon for his own name, an arrangement to be achieved only by displacing all the names used by Hevelius at risk of considerable perplexity and confusion to later astronomers." *Der Mond* (Berlin, 1837), pp. 184–185. This may be true, but if one considers a few of Hevelius's names in their Latin form—*Lacus Hyperboreus superior, Lacus Hyperboreus inferior, Lacus Corocondametis*, and *Promontorium Leucopetria*, for example—one is inclined somewhat to forgive Riccioli for his vanity.

7. E. J. Stone, "Note on a Crayon Drawing of the Moon by John Russell, R. A., at the Radcliffe Observatory, Oxford," *Monthly Notices of the Royal Astronomical Society*, 56 (1896), 92.

8. Owen Gingerich, "Galileo and the Phases of Venus," *Sky and Telescope*, 68 (1984), 520–522.

9. Kepler, *Conversation*, p. 37.

10. Giordano Bruno, *De innumerabilis, immenso, et infigurabilis* (Frankfurt, 1591).

11. Kepler, *Conversation*, p. 44.

12. Galileo Galilei, *The Sidereal Messenger*, trans. E. S. Carlos, 1880, in *A Source Book in Astronomy* ed. Harlow Shapley and Helen E. Howarth (New York, 1929), p. 47; Milton, *Paradise Lost*, bk. VII, line 581.

13. Kepler, *Conversation*, p. 40.

14. Tycho's opinion is cited by Kepler in his *Harmonice Mundi* (Harmonies of the World), bk. V; a translation may be found in Great Books of the Western World, ed. Robert Maynard Hutchins (Chicago: Encyclopaedia Britannica, 1952), vol. 16, pp. 1084–1085.

15. Lucretius, *De Rerum Natura, Libri Sex*, trans. Cyril Bailey (Oxford, 1966), bk. II, lines 1052–1057.

16. Ibid., bk. II, lines 1074–1076.

17. See Paul Stroobant, "Etude sur le satellite énigmatique de Vénus," *Astronomische Nachrichten*, no. 2809 (1887), for the classic discussion of the Venus satellite observations.

18. Fontenelle, *Entretiens*, p. 54.

19. The full title is *Kosmotheoros, sive Coelestibus, earumque ornatu, conjecturae* (The Hague, 1698). The English translation, by John Glanville, was published in London in 1722.

## Chapter 4.
## To Complete the Analogy

1. Quoted in Richard A. Proctor, *The Moon: Her Motions, Aspect, Scenery, and Physical Condition*, 2nd ed. (London, 1878), p. 184.

2. Ibid.

3. In general, a telescope magnifying $M$ times will make the Moon or a planet have the apparent size it would have as seen with the naked eye at a distance $D' = 1/M \times D$, where $D$ is the actual distance from the Earth.

4. Webb, *Celestial Objects*, vol. 1, p. 78. Webb adds: "The existence of many natural wonders on our own globe—for example, the cañons, or river-gorges, of NW America, one of which has a length of 550 miles, an extreme depth of 7000 feet, and a closest contraction of 100 feet, or the obelisk of limestone, near Lanslebourg, 360 feet high with a base of 40 feet . . . shows how cautiously inferences should be drawn as to the artificial origin of extraordinary appearances" (p. 79n).

5. Agnes M. Clerke, *A Popular History of Astronomy During the Nineteenth Century*, 3d ed. (London, 1893), p. 300.

6. Johann Hieronymus Schröter, *Selenotopographische Fragmente zur genauern Kentniss der Mondfläche* (Lilienthal, Privately Published, 1st ser., 1791; 2d ser., 1802).

7. Wilhelm Beer and Johann Heinrich von Mädler, *Der Mond* (Berlin, 1837), p. 185.

8. Philipp Fauth, *The Moon*, trans. Joseph McCabe (New York, 1909), pp. 30–31. Fauth, by the way, supported the strange "world ice doctrine" of Hanns Hörbiger, and Schröter's ideas seem mild by comparison.

9. Webb, *Celestial Objects*, vol. 1, p. 106.

10. Sir William Herschel, "An account of three volcanoes on the moon," *Philosophical Transactions of the Royal Society*, 77 (1787), 229–232 (hereafter cited as *Phil. Trans.*); and Herschel, "On the nature and construction of the sun and fixed stars," *Phil. Trans.*, 85 (1795), 46–72.

11. Franz von Paula Gruithuisen, *Entdeckung vieler deutlichen Spüren der Mondbewohner* (Discovery of Many Significant Traces of the Inhabitants of the Moon) (Nürnberg, 1824). The very first to propose the meteoritic theory seems to have been Marschall von Bieberstein in 1802.

12. Quoted in Webb, *Celestial Objects*, vol. 1, p. 87n.

13. Ibid., p. 120.

14. Beer and Mädler, *Der Mond*, p. 287.

15. The full title is *Der Mond: Nach seinem kosmischen und individuellen verhältnissen; oder, Allgemeine vergleichende selenographie* (The Moon: Concerning Its Cosmic and Individual Conditions; or, A Comparative Study of General Selenography). The book runs to over 400 folio pages. If there is any truth to a comment William James once made to Percival Lowell concerning Darwin—that "his greatness was due to his great detail as increasing the probabilites; showing again how mere detail, mere bulk impresses, though probability be not furthered a bit"— the same certainly applies to Beer and Mädler's chef d'oeuvre.

16. Webb, *Celestial Objects*, vol. 1, p. 88.

17. Beer and Mädler, *Der Mond*, p. 134.

18. Alfred, Lord Tennyson, "Locksley Hall Sixty Years After," line 175.

19. Richard A. Proctor, *The Moon*, pp. 183–184. Compare Edmund Neison, whose revision of Beer and Mädler's map appeared in 1876, two years before Proctor wrote these words. Beer and Mädler's work, Neison says, gave the impression that "the moon was to all intents an airless, waterless, lifeless, unchangeable desert, with its surface broken by vast extinct volcanoes. With this opinion prevailing the natural effect of such great works as Beer and Mädler's speedily ensued, the attention of astronomers was directed to other fields, and Selenography resting on its laurels made no further progress for many years." Edmund Neison, *The Moon: And the Condition and Configuration of its Surface* (London, 1876), pp. 104–105.

20. Julius Schmidt "Ueber den Mondcrater Linné," *Astronomische Nachrichten*, no. 1631 (1867).

21. See Richard J. Pike, "The Lunar Crater Linné," *Sky and Telescope*, 46 (1973), 364–366.

22. Webb, *Celestial Objects*, vol. 1, p. 78.

23. Quoted in Myron A. Hoffer, *The Roots of Human Behavior* (San Francisco, 1981), p. 7.

24. Gombrich, *Art and Illusion*, p. 74.

25. Percival Lowell, *The Soul of the Far East* (New York, 1888), p. 209.

26. Quoted in Willy Ley, *Watchers of the Skies: An Informal History of Astronomy from Babylon to the Space Age* (New York, 1963), p. 487.

27. Webb, *Celestial Objects*, vol. 1, pp. 71–72.

28. William Herschel, "On the remarkable appearances at the polar regions of the planet *Mars*, the inclination of its axis, the position of its poles and its spheroidical figure; with a few hints relating to its real diameter and atmosphere," *Phil. Trans.*, 74 (1784), 233–273.

29. Quoted in Webb, *Celestial Objects*, vol. 1, p. 169n.

## Chapter 5.
## Like a Distant View of Earth

1. Tennyson, "Locksley Hall Sixty Years After," lines 183–192.

2. Giovanni Domenico Cassini, "Lettre à M. Petit touchant la découverte du mouvement de la Planète Venus autour de son axe," *Journal des Scavans* (Paris, 1667), p. 122. Through a typographical error, the sentence in question reads "Less than a day, namely 23 days."

3. F. Bianchini, *Hesperi et Phosphori nova phaenomena* (Rome, 1727). Jacques Cassini, Giovanni Domenico's son, later attempted to reconcile this result with that arrived at by his father. He pointed out that if Bianchini had observed the planet after it had made *several* rotations, rather than only one, he could have derived a period of 23 hours, 20 minutes.

4. J. H. Schröter, "Observations on the Atmosphere of Venus and the Moon, their respective Densities, perpendicular Heights, and the Twilight occasioned by them," *Phil. Trans.*, 82 (1792), 337.

5. W. Herschel, "Observations on the Planet Venus," *Phil. Trans.*, 83 (1793), 201.

6. J. H. Schröter, "New observations on further Proof of the mountainous Inequalities, Rotation, Atmosphere, and Twilight, of the Planet Venus," *Phil. Trans.*, 85 (1795), 117.

7. An interesting photograph of such an irregularity, by Alan W. Heath, appears in the *Journal of the Association of Lunar and Planetary Observers*, 31 (1985), 82.

8. J. H. Schröter, "Atmospheres of Venus and the Moon," 309; also *Aphroditographische Fragmente* (Erfurt, 1796), p. 85.

9. Webb, *Celestial Objects*, vol. 1, p. 65.

10. F. De Vico, "Beobachtungen von Flecken auf der Venus im Collegio Romano," *Astronomische Nachrichten*, no. 404 (1840). A similar appearance was, however, recorded forty years later by W. F. Denning, an observer of undoubted skill (as noted in Clerke, *Popular History of Astronomy*, p. 311).

11. The phrase is from the *Aeneid*, bk. VI, line 454. The full context, from the Loeb translation, is as follows: "Among them, with wound still fresh, Phoenician Dido was wandering in the great forest, and as soon as the Trojan hero stood nigh and knew her, a dim form amid the shadows—even as, in the early month, one sees, or fancies he has seen the moon rise amid the clouds—he shed tears and spoke to her in tender love."

12. Percival Lowell, *Mars* (Boston, 1895), p. 21.

13. William Herschel, *Phil. Trans.*, 74 (1784), 260.

14. W. Beer and J. H. von Mädler, *Physikalisches Beobachtungen des Mars in der Erdnähe* (Berlin, 1830).

15. J. H. Schröter, *Areographische Beiträge zur genaueren Kenntnis und Beurteilung des Planeten Mars* (Leyden, 1881).

16. Ley, *Watchers of the Skies*, p. 290.

17. Webb, *Celestial Objects*, vol. 1, pp. 172–173.

18. Ibid., pp. 175–176.

19. The map first appeared in Proctor's book *Other Worlds than Ours* (London, 1870).

20. Joseph Ashbrook, "The Eagle Eye of William Rutter Dawes," in *Sky and Telescope*, 46 (1973), 27–28, is the source for Dawes's life mainly followed here.

21. Ibid., p. 27.

22. Christiaan Huygens, *Treatise on Light* (The Hague, 1690).

23. For information concerning the patterns produced by other than circular lenses, such as apertures with a central obstruction (which is the case for reflectors), see David E. Stoltzmann, "Resolution Criteria for Diffraction-Limited Telescopes," *Sky and Telescope*, 65 (1983), 176–181.

24. George Biddell Airy, *Transactions of the Cambridge Philosophical Society*, 5 (1835), 283. For Airy's role in the story of the discovery of Neptune, see Morton Grosser, *The Discovery of Neptune* (New York, 1979).

25. Rev. William Rutter Dawes, *Memoirs of the Royal Astronomical Society*, 35 (1865), 154.

*Chapter 6.*
*Satellites and Seeing*

*285*
———
*Notes to*
*Pages*
*56–62*

1. Because the orbit of Mars is tilted by 1°51' to that of the Earth, the date of closest approach actually preceded that of opposition by three days. The minimum distance attained was 34,992,000 miles (56,315,000 kilometers).

2. A source for the life of Asaph Hall is Joseph Ashbrook, "Asaph Hall Finds the Moons of Mars," *Sky and Telescope*, 54 (1977), 20–21.

3. Asaph Hall, "My Connection with the Harvard Observatory and the Bonds—1857–1862," in Edward Singleton Holden, *Memoirs of William Cranch Bond and of his son George Phillips Bond* (San Francisco, 1897), pp. 77–78.

4. For the history of the Harvard College Observatory, see Bessie Zaban Jones and Lyle Gifford Boyd, *The Harvard College Observatory: The First Four Directorships, 1839–1919* (Cambridge, Mass., 1971).

5. This telescope had originally been ordered by the University of Mississippi at Oxford, but with the outbreak of the Civil War it could not be delivered. It was set up instead at the old University of Chicago, and it was later moved to nearby Evanston, where it is today.

6. Quoted in Jones and Boyd, *Harvard College Observatory*, p. 454n. For details of Clark's life, see Deborah Jean Warner, *Alvan Clark and Sons: Artists in Optics* (Washington, D.C., 1968).

7. Alvan Clark, "Autobiography of Alvan Clark," *Sidereal Messenger*, 8 (1889), 113.

8. "Discovery of a companion of Sirius," *Monthly Notices of the Royal Astronomical Society*, 22 (1862), 170.

9. From the report of Admiral Benjamin F. Sands, superintendent of the U.S. Naval Observatory, Oct. 10, 1868, quoted in R. W. Rhynsburger, "A Historic Refractor's 100th Anniversary," *Sky and Telescope*, 46 (1973), 208.

10. Ibid.

11. Asaph Hall, "The Period of Saturn's Rotation," *Monthly Notices of the Royal Astronomical Society*, 38 (1878), 209; Asaph Hall, "The Satellites of Mars," in *A Source Book in Astronomy* (1878), p. 320. See also a letter from Asaph Hall to J. W. L. Glaisher of Dec. 28, 1877, published as "The Discovery of the Satellites of Mars," *Monthly Notices of the Royal Astronomical Society*, 38 (1878), 205. In this letter Hall describes the effect that his discovery of the rotation of Saturn had on leading him to undertake the quest for the Martian satellites. The wording of the letter is slightly different from, but follows closely, the account of the discovery of the satellites in Hall's article "The Satellites of Mars."

12. See Owen Gingerich, "The Satellites of Mars: Prediction and Discovery," *Journal for the History of Astronomy*, 1 (1970), 109–115.

13. Hall, "Satellites of Mars," p. 321.

14. Ibid.

15. *The Iliad of Homer*, trans. Richmond Lattimore (Chicago, 1951), bk. XV, lines 119–120.

16. In addition to the case of Deimos, Saturn's satellite Mimas, discovered by W. Herschel with an 18-inch reflector, was later reported in a 6¹/₃-inch glass, and Enceladus with still smaller instruments. The two outer satellites of Uranus, discovered by Herschel with the same instrument, were reported in a 6¹/₂ inch-reflector; and Neptune's Triton, first grasped by W. Lassell with a 24-inch reflector, was afterwards seen by Dawes with an 8-inch refractor. This list does not include the almost unbelievable claims of the nineteenth-century Irish observer I. W. Ward, of which more will be said in Chapter 9.

17. Webb, *Celestial Objects,* vol. 1, p. 213.

18. Ibid., p. 212.

19. Quoted in Norwood Russell Hanson, *Patterns of Discovery* (Cambridge, 1959), p. 184.

20. Quoted in Joseph Ashbrook, "The Eagle Eye of William Rutter Dawes," *Sky and Telescope,* 46 (1973), 27–28.

21. Pierre Simon de Laplace, "Mémoire sur la theorie de l'anneau de Saturne," *Mémoires de l'Academie royale des Sciences de Paris* (1789), reprinted in *Oeuvres complètes de Laplace* (Paris, 1878–1912), vol. 11, pp. 275–292. See also Laplace's *Mécanique Céleste* (Paris, 1799), vol. 2, bk. III, chap. 6.

22. James Clerk Maxwell, *On the Stability of the Motion of Saturn's Rings* (Cambridge, 1859). Reprinted, with additional material, in *Maxwell on Saturn's Rings,* ed. Stephen G. Brush, C.W.F. Everitt, and Elizabeth Garber, 69–157 (Cambridge, Mass., 1983).

23. Henry Kater, "On the Appearance of Divisions in the Exterior Ring of Saturn," *Memoirs of the Royal Astronomical Society,* 4 (1831), 383–390. Kater notes that two friends also observed the planet that night. One saw six divisions, and the other, who was "extremely short-sighted" and "not much accustomed to telescopic observations," could only make out one.

24. L. A. J. Quételet is mentioned in Kater, "Exterior Ring of Saturn," p. 388. For J. F. Encke, see his article "Über den Ring des Saturns," in *Mathematische Abhandlungen der Königlichen Akademie der Wissenschaften zu Berlin, Aus dem Jahre 1838* (1840), 1–18. His drawing is reproduced in Donald E. Osterbrock and Dale P. Cruikshank, "J. E. Keeler's Discovery of a Gap in the Outer Part of the A Ring," *Icarus,* 53 (1983), 167.

25. G. P. Bond, "On the Rings of Saturn," *American Journal of Science,* 2 (1851), 97–105; and *Astronomical Journal,* 2 (1851), 5–8, 10.

26. Harvard College Observatory *Annals,* vol. 2, p. 50n.

27. Maunder, "'Eye of Mars,'" p. 57.

28. William James, *The Principles of Psychology* (1890; reprint, New York, 1950), vol. 1, p. 425.

29. Ibid., p. 428.

30. Ibid., p. 439.

31. Ibid., p. 443.

Chapter 7.
Schiaparelli

287

———

Notes to
Pages
68–73

1. Schiaparelli's work on the precursors of Copernicus, and especially on the planetary theory of Aristarchus, is noteworthy. He also wrote on the astronomy of the Old Testament, concluding that the Hebrew word *Mazzaroth,* which appears in Job 38:32, probably referred to the planet Venus. But probably his greatest contribution was his detailed working out of the theory of the homocentric spheres of Eudoxus of Cnidus, for which see "Le sfere omocentriche di Eudosso, di Callippo e di Aristotele," *Publicazione del Royal Osservatorio di Brera in Milano,* no. 9 (Milan, 1875).

2. For Schiaparelli's career, necrological sources include E. B. Knobel, obit. note, *Monthly Notices of the Royal Astronomical Society,* 71 (1911), 282–287; Hector MacPherson, "Giovanni Schiaparelli," *Popular Astronomy,* 18 (1910), 467–474 (hereafter cited as *Pop. Ast.*); and Camille Flammarion, "L'Astronome Schiaparelli," *Bulletin de la Société Astronomique de France,* 24 (1910), 375–377 (hereafter cited as *Bull. Soc. Astronomique de France*). Other useful sources include Hector MacPherson, *Makers of Astronomy* (Oxford, 1933), pp. 189–194, and Giovanni Cossavella, *L'Astronomo Giovanni Schiaparelli* (Turin, 1914).

3. MacPherson, *Makers of Astronomy,* p. 189.

4. MacPherson, "Giovanni Schiaparelli," p. 468.

5. G. V. Schiaparelli, "Intorno alla Teoria Astronomica della . . . Stelle Cadenti," *Memoire Società Italiana della Scienza,* ser. 3, vol. 1, pt. 1 (1867), in *Le Opere di G. V. Schiaparelli* (1930; reprint, New York, 1969), vol. 3, pp. 337–451.

6. As noted in MacPherson, *Makers of Astronomy,* p. 190.

7. G. V. Schiaparelli, "Intorno al corso ed all'origine probabile delle Stella Meteoriche: Lettere al P. A. Secchi," *Bulletino Meteorologico dell'Osservatòrio del Collegio Romano,* vols. 5, 6 (1866–67), in *Opere,* vol. 3, pp. 261–316.

8. As he told E. E. Barnard. See George van Biesbroeck, "E. E. Barnard's Visit with Giovanni Schiaparelli," *Pop. Ast.,* 42 (1934), 553–558.

9. G. V. Schiaparelli, "Considerazioni sul Moto Rotatorio del Pianeta Venere" (1890), in *Opere,* vol. 5, pp. 400–410.

10. G. V. Schiaparelli, "Sulla Rotazione di Mercurio," *Astronomische Nachrichten,* no. 2944 (1889), reprinted in *Opere,* vol. 5, pp. 325–335. He told Barnard in 1893: "For Mercury I have not the slightest doubt. For Venus the observations are too scarce and somewhat doubtful: the rotation I regard as probable but not definite." George van Biesbroeck, "E. E. Barnard's Visit with Giovanni Schiaparelli," *Pop. Ast.,* 42 (1934), p. 554.

11. G. V. Schiaparelli, "Sulla Rotazione e Sulla Costituzione Fisica del Pianeta Mercurio," an address to the Royal Academy of Lynxes, Dec. 8, 1889, reprinted in *Opere,* vol. 5, p. 344.

12. Knobel, Schiaparelli obit. note, p. 286.

13. William Herschel, "On the ring of Saturn, and the rotation of the fifth satellite upon its axis," *Phil. Trans.*, 82 (1792), 1–22.

14. According to E. M. Antoniadi, *The Planet Mercury,* trans. Patrick Moore (Devon, 1974), p. 28.

15. Ibid.

16. Ibid., p. 59.

17. Ibid.

18. Ibid., pp. 61–62.

19. Secchi is quoted in E. M. Antoniadi, "La Vie dans L'univers," *Bull. Soc. Astronomique de France,* 52 (1938), 6.

20. Schiaparelli, Address to the Royal Academy of Lynxes, quoted in *Watchers of the Skies,* p. 192.

21. Percival Lowell, "Mercury," *Monthly Notices of the Royal Astronomical Society,* 57 (1896–97), 148.

22. Percival Lowell, *The Evolution of Worlds* (New York, 1910), p. 68.

23. Ibid., p. 71.

24. Rev. T.E.R. Phillips, W. F. Denning, et al., *The Splendour of the Heavens* (London, 1928), p. 192.

25. Antoniadi, *The Planet Mercury,* pp. 30–31.

26. Ibid., p. 31.

27. Ibid., pp. 37–38.

28. Ibid., p. 63.

29. As noted in Werner Sandner, *The Planet Mercury* (New York, 1963), p. 30.

30. Dale P. Cruikshank and Clark R. Chapman, "Mercury's Rotation and Visual Observations," *Sky and Telescope,* 34 (1967), 24–26. See also the excellent account in Clark R. Chapman, *Planets of Rock and Ice: From Mercury to the Moons of Saturn* (New York, 1982), pp. 63–72.

31. Schiaparelli, "Sulla Rotazione de Mercurio," *Opere,* vol. 5, p. 333.

32. Ibid., pp. 333–334.

33. Schiaparelli, "Sulla Rotazione e Sulla Costituzione del Pianeta Mercurio," p. 344.

34. E. Walter Maunder, "The Canals of Mars," *Knowledge,* Nov. 1, 1894, p. 252.

35. Cruikshank and Chapman, "Mercury's Rotation," p. 25.

36. G. V. Schiaparelli, "Il Pianeta Marte," in *Opere,* vol. 2, pp. 47–74; "The Planet Mars," trans. W. H. Pickering, *Astronomy and Astro-Physics,* 13 (1894), pp. 635–640, 714–723 (quotations are from p. 719; I have made use of Pickering's translation throughout); Percival Lowell, *Mars* (Boston, 1895), p. 153.

37. Schiaparelli, "The Planet Mars," pp. 721–722; Schiaparelli, quoted in Ley, *Watchers of the Skies,* p. 192.

38. Schiaparelli, "The Planet Mars," p. 723.

39. Quoted in Hoyt, *Lowell and Mars,* p. 90.

40. Michael J. Crowe makes the case very convincingly in his extremely thorough book *The Extraterrestrial Life Debate: 1750–1910* (Cambridge, 1986). Professor Crowe was kind enough to allow me to

read the chapter of his work that was concerned with Mars while the book was still in manuscript; his ideas have had a considerable influence on my own, especially regarding Schiaparelli.

41. Rev. T. W. Webb, "Planets of the Season: Mars," *Nature*, 21 (1880), 213.

42. G. V. Schiaparelli, "Osservazioni Astronomiche e Fisiche sull'Asse di Rotazione e sulla Topografia del Pianeta Marte . . . durante l'opposizione del 1877–MEMORIA PRIMA," in *Opere*, vol. 1, p. 60.

43. Quoted in Beer and Mädler, *Der Mond*, p. 190.

44. P. Lowell, *Mars*, p. 141.

45. Schiaparelli, "The Planet Mars," p. 717.

46. Schiaparelli, "Auf der Oberfläche des Planeten Mars," *Opere*, vol. 2, p. 23.

47. P. Lowell, *Mars*, p. 157.

48. The account that follows draws upon G. V. Schiaparelli, "Osservazioni del Pianeta Marte," pp. 66–69, and "Auf der Oberfläche des Planeten Mars," p. 36.

49. Vincenzo Cerulli, *Marte nel 1896–1897* (Collurania, It. 1898), p. 115, quoted in E. M. Antoniadi, "Mars Section, Fifth Interim Report for 1909," *Journal of the British Astronomical Association*, 20 (1909), 139 (hereafter cited as *Journal of the BAA*).

50. E. Walter Maunder, "The Canals of Mars," *Knowledge*, Nov. 1, 1894, p. 250: "The 'canals,' when near the edge of the disc, are apt to be represented as much straighter than they could possibly be."

51. E. M. Antoniadi, "Le Retour de la Planète Mars," *Bull. Soc. Astronomique de France*, 40 (1926), 350.

52. Quoted in W. L. Leonard, *Percival Lowell: An Afterglow* (Boston, 1921), p. 67.

53. Webb, "Planets of the Season: Mars," p. 212.

## Chapter 8.
### The Art of Observing

1. The account of Draper's life and of his photographs of the Orion Nebula which follows owes much to Owen Gingerich, "The First Photograph of the Orion Nebula," *Sky and Telescope*, 60 (1980), 364–366; and to Edward Singleton Holden, "Monograph of the Central Parts of the Nebula of Orion," *Washington Astronomical Observations for 1878* (Washington, D.C., 1882), Appendix I, Addendum, pp. 226–230.

2. Gingerich, "Orion Nebula," p. 364.

3. Asaph Hall, "My Connection with the Harvard Observatory and the Bonds—1857–62," in Edward Singleton Holden, *Memorials of William Cranch Bond and of His Son George Phillips Bond* (San Francisco, 1897), pp. 78–79.

4. Bond's drawing, by the way, does an excellent job of capturing that appearance of the central nebula which is hardly less vividly depicted in the words of Sir John Herschel: "I know not how to describe it better

than by comparing it to a curdling liquid, or to the breaking up of a mackerel sky." John Herschel, "Account of Some Observations with a 20-foot Reflecting Telescope," *Memoirs of the Royal Astronomical Society*, 2 (1826), 491.

5. The analogy is, in fact, rather precise. See J. Rösch, G. Courtès, and J. Dommanget, "Le Choix des Sites d'Observatoires Astronomiques," *Bulletin Astronomique de l'Observatoire de Paris*, 24 (1963), fasc. 2, 3, which contains images from a motion picture of lunar detail and of the double star 85 Peg taken through the 24-inch refractor at the Lowell Observatory. "Good frames were post-selected and showed that the full resolving power of the full aperture was obtained for occasional frames out of thousands. Most of the movies were taken at 30 to 60 frames per second, so the eye ought to be able to appreciate such instants of best image resolution." Arthur A. Hoag first called my attention to this very interesting paper and provided the comments on it quoted here.

6. Floyd Ratliff, "The Role of Physiological Nystagmus in Monocular Acuity," *Journal of Experimental Psychology*, 43 (1952), 163–172.

7. Ülker T. Keesey, "Effects of Involuntary Eye Movements on Visual Acuity," *Journal of the Optical Society of America*, 50 (1960), 769–774.

8. G. V. Schiaparelli, "Observations of the Planet Mars" [review of the first volume of the Lowell Observatory *Annals*], *Science*, May 5, 1890; reprinted in *Opere*, vol. 2, p. 252.

9. G. V. Schiaparelli, "Osservazioni astronomiche e fisiche del Pianeta Marte," p. 73.

10. Samuel Johnson, "Milton," in *The Works of Samuel Johnson, LL.D.*, ed. Arthur Murphy (London, 1810), vol. 9, p. 109.

11. Francis Bacon, *The Advancement of Learning, book I*, ed. William A. Armstrong (London, 1975), pp. 79–80.

12. G. V. Schiaparelli, "On Some Observations of Saturn and Mars," *Observatory*, 5 (1882), 224, 223.

13. G. V. Schiaparelli to N. E. Green, October 27, 1879, cited in a note by Green in "Mars and the Schiaparelli Canals," *Observatory*, 3 (1879), 252.

14. N. E. Green, "Observations of Mars, at Madeira in Aug. and Sept. 1877," *Memoirs of the Royal Astronomical Society*, 44 (1877–79), 123–140.

15. Green, "Mars and the Schiaparelli Canals," p. 252.

16. Ibid.

17. Green, "On Some Changes in the Markings of Mars, since the Opposition of 1877," *Monthly Notices of the Royal Astronomical Society*, 40 (1880), 332.

18. "Proceedings of the Meeting of the Royal Astronomical Society, April 14, 1882," *Observatory*, 5 (1882), 135–137.

19. "Report of the Meeting of the Association Held Dec. 31, 1890," *Journal of the BAA*, 1 (1890), 112.

20. Webb, "Planets of the Season: Mars," p. 213.

21. Ibid.

22. Richard A. Proctor, "Maps and Views of Mars," *Scientific American Supplement*, 26 (1888), 10659–10660.

23. C. A. Young, "Observations of Mars at the Halsted Observatory, Princeton, New Jersey," *Astronomy and Astro-Physics*, 11 (1892), 675–678.

24. As reported in "Proceedings at the Meeting of the Royal Astronomical Society," *Observatory*, 16 (1893), 218.

25. Ibid.

26. Ibid.

27. George Van Biesbroeck, "E. E. Barnard's Visit with Giovanni Schiaparelli," *Pop. Ast.*, 42 (1934), 553–558.

28. G. V. Schiaparelli, "Observations of the Planet Mars," *Opere*, vol. 2, pp. 251, 249.

29. William James, *Principles of Psychology* (1890; reprint, New York, 1950), vol. 2, p. 598.

30. W. W. Campbell, Review of *Mars*, by Percival Lowell, *Publications of the Astronomical Society of the Pacific*, 8 (1896), 208–220.

31. E. S. Holden, *The Forum*, Nov. 1892.

32. E. M. Antoniadi, "Report of Mars Section, 1909," *Memoirs of the British Astronomical Association*, 20 (1916), 38 (hereafter cited as *Memoirs of the BAA*).

33. E. M. Antoniadi, "Mars Section, Fifth Interim Report, 1909," *Journal of the BAA*, 20 (1909), 141.

## Chapter 9.
## Of Aperture and Atmosphere

1. W. F. Denning, "Jupiter and the Relative Powers of Telescopes in Defining Planetary Markings," *Observatory*, 8 (1885), 80. For the chart by Harkness and Hall, see *Monthly Notices of the Royal Astronomical Society*, 40 (1879), 13.

2. William Noble, "Anomalies of Vision," *Astronomical Register*, 15 (1877), 222–225.

3. Ibid.

4. Herbert Sadler, "Miraculous Vision," *Astronomical Register*, 15 (1877), 298–301.

5. Ibid.

6. W. F. Denning, "Large vs. Small Telescopes," *Observatory*, 9, (1886), 276.

7. Jan K. Herman, *A Hilltop on Foggy Bottom* (Washington, D. C., 1984), p. 52.

8. W. F. Denning, *Telescopic Work for Starlight Evenings* (London, 1891), p. 25.

9. Sir Isaac Newton, *Opticks*, 4th ed. (London, 1730), bk. I, pt. 1, prop. viii, prob. 2.

10. William Lassell, quoted in W. F. Denning, "Large vs. Small Telescopes," p. 276.

11. Lord Rosse, Appendix to *Royal Dublin Society Scientific Transactions*, Aug. 1879, quoted in W. F. Denning, "Jupiter and the Relative Powers of Telescopes in Defining Planetary Markings," p. 79.

12. Ibid.

13. G. B. Airy, "Notice of the Substance of a Lecture on the Large Telescopes of the Earl of Rosse and Mr. Lassell," *Monthly Notices of the Royal Astronomical Society,* 9 (1849), 120.

14. W. F. Denning, "The Defining Power of Telescopes," *Observatory,* 8 (1885), 209.

15. Obit. notice by A.G.C. Crommelin, *Journal of the BAA,* 42 (1931), 81.

16. Denning, "Defining Power of Telescopes," p. 209.

17. Denning, "Jupiter and the Relative Powers of Telescopes," pp. 79–80.

18. Asaph Hall, "The Defining Power of Telescopes," *Observatory,* 8 (1885), 174.

19. Quoted in Denning, "Defining Power of Telescopes," p. 206.

20. Ibid.

21. C. A. Young, "Large Telescopes vs. Small," *Observatory,* 9 (1886), 92.

22. C. A. Young, "Small Telescopes vs. Large," *Sidereal Messenger,* 5 (1886), 1–5.

23. Harvard College Observatory *Annals,* vol. 2, pt. 1 (Cambridge, 1857). In the introduction (p. 2) there appear the following comments: "Although in a few instances powers above 1,000 have been applied, as a general rule those between 300 and 700 have been preferred on occasions of fine definition. So many nights are rendered unavailing for the purposes of nice observation, from the annoyance of atmospheric disturbances, that constant watchfulness and the outlay of much fruitless labor are called for to secure only a moderate number of views in which the qualities of the instrument can be adequately appreciated."

24. C. A. Young, "Observation of the Red Spot of Jupiter," *Observatory,* 8 (1885), 172–174.

25. Denning, "Large vs. Small Telescopes," p. 277.

26. Young, "Large Telescopes vs. Small," p. 94.

27. Young, "Small Telescopes vs. Large," p. 2.

*Chapter 10.
Confirmations?*

1. Webb, "Planets of the Season: Mars," p. 213.

2. Quoted in T.J.J. See, "The Study of Planetary Detail," *Pop. Ast.,* 4 (1897), 553.

3. E. M. Antoniadi, "Report of Mars Section, 1898–1899," *Memoirs of the BAA,* 9 (1901), 68.

4. G. V. Schiaparelli to Hector MacPherson, Nov. 16, 1903, quoted in "Giovanni Schiaparelli," *Pop. Ast.,* 18 (1910), 474.

5. G. V. Schiaparelli, "Dedica della Memoria Terza Sopra Marte al Prof. Tito Vignoli," in *Opere,* vol. 2, pp. 477–478. Michael Armstrong assisted me by providing a literal translation of the Latin; the versifica-

tion is my own. The poem was signed Achilles Parius, an anagram on the name Schiaparellius.

6. François Terby, "Physical Observations of Mars," *Astronomy and Astro-Physics*, 11 (1892), 479–480.

7. P. Lowell, *Mars*, p. 160.

8. William Herschel, quoted in Webb, *Celestial Objects*, vol. 1, p. xii.

9. E. H. Gombrich, *Art and Illusion*, p. 204.

10. See R. H. Austin, "Uranus Observed," *British Journal for the History of Science*, 3 (1967), 275–284. Austin's conclusion is that the only possible explanation is autosuggestion.

11. Agnes M. Clerke, *Popular History of Astronomy*, p. 483.

12. P. Lowell, "Schiaparelli," *Pop. Ast.*, 18 (1910), 462.

13. P. Lowell, *Mars*, p. 140. Not so very early, though: Goethe's Mephistophilis (1808) is the "Geists der stets verneint" (spirit that denies).

14. Unsigned note in *Observatory*, 9 (1886), 364.

15. P. Lowell, *Mars*, p. 138.

16. Unfortunately, its awkward length made it mandatory to have two people to work the telescope, and as the Nice Observatory had a rule, "one man, one telescope," it soon fell into relative disuse, so Perrotin's observations seem to have been not only the first but also virtually the last observations made with it of any scientific importance.

17. For the early history of the Lick Observatory, see Donald E. Osterbrock, "The California-Wisconsin Axis in Astronomy, I," *Sky and Telescope*, 51 (1976), 9–14; also Michael Chriss, "The Stars Move West: The Founding of Lick Observatory," *Mercury*, 2 (1973), 10–15.

18. "The will of James Lick," in *A Source Book in Astronomy*, ed. Shapley and Howarth, p. 316.

19. J. E. Keeler, "The Outer Ring of Saturn," *Astronomical Journal*, 8 (1889), 175. See also Osterbrock and Cruikshank, "J. E. Keeler's Discovery," pp. 165–173.

20. J. E. Keeler, "First Observations of Saturn with the 36-inch Equatorial of Lick Observatory," *Sidereal Messenger*, 7 (1888), 79–83.

21. Camille Flammarion, *La Planète Mars et ses conditions d'habitabilité* (Paris, 1892), pp. 426–430.

22. Quoted in MacPherson, "Giovanni Schiaparelli," p. 471.

23. Schiaparelli to François Terby, June 8, 1888, quoted in E. W. Maunder, "The Canals of Mars," *Observatory*, 11 (1888), 347–348.

24. E. S. Holden, J. M. Schaeberle, and J. E. Keeler, "Note on the Opposition of Mars, 1890," *Publications of the Astronomical Society of the Pacific*, 2 (1890), 299–300.

25. Young, "Observations of Mars," pp. 675–678.

26. Ibid., pp. 676–677.

27. E. S. Holden, "Note on the Mount Hamilton Observations of Mars, June–August, 1892," *Astronomy and Astro-Physics*, 11 (1892), 663.

28. Clerke, *Popular History of Astronomy*, p. 344.

29. Ibid.

30. Ibid., p. 341.

1. For details, see Dale P. Cruikshank, "Barnard's Satellite of Jupiter," *Sky and Telescope*, 64 (1982), 221–224.

2. Biographical information on Barnard can be found in E. B. Frost, "Edward Emerson Barnard," *Astrophysical Journal*, 58 (1923), 1–35; Phillip Fox, "Edward Emerson Barnard," *Pop. Ast.*, 71 (1923), 195–200; S. W. Burnham, "Early Life of E. E. Barnard," *Pop. Ast.*, 1 (1894), 193–195, 341–345, 441–447; and Edward Emerson Barnard, "Some Experiences as an Astronomer," *Christian Advocate* (Nashville, Tenn.), July 5, 1907, pp. 23–28.

3. John E. Mellish's letter to Walter Leight, dated January 18, 1935, was first published in Rodger W. Gordon, "Mellish and Barnard—They Did See Martian Craters!" *Strolling Astronomer* 25 (1975), 196. Mellish's first published account of his observations did not appear until after the Mariner 4 flyby; see his letter to the editor in *Sky and Telescope*, 31 (1966), 339.

4. Gordon, "Mellish and Barnard," p. 196.

5. Mellish, letter to the editor, p. 339.

6. Barnard's preliminary report was published in *Astronomy and Astro-Physics*, 11 (1892), 680–684. For his meeting with Schiaparelli, see George van Biesbroeck, "E. E. Barnard's Visit with Giovanni Schiaparelli," *Pop. Ast.*, 42 (1934), 553–555. Barnard's use of only 350× on the Lick refractor shocked Schiaparelli, who regularly used 420× on his 8.6-inch refractor and 650× on his 18-inch, which he had begun using in 1886. He told Barnard: "Here, where the weather is favorable, the amplifications 200 and 350 are not enough to do justice to the instrument." However, Antoniadi, in his observations of Mars with the 32³/₄-inch refractor at Meudon, also used relatively low powers, because more than 320× was, he found, "already too much for the delicate shadings." See E. M. Antoniadi, "Report of Mars Section, 1909," *Memoirs of the BAA*, 20 (1916), 28. Schiaparelli, incidentally, found seeing conditions in Milan best "with the east wind, with moderate velocity"; van Biesbroeck, p. 554.

7. Quoted in W. H. Hartmann and O. R. Roper, *The New Mars: The Discoveries of Mariner 9* (Washington, D.C., 1974), pp. 6–7.

8. E. E. Barnard, "Micrometric Measures of the Ball and Ring System of the Planet Saturn . . . with some remarks on large and small telescopes," *Monthly Notices of the Royal Astronomical Society*, 56 (1896), 163–164.

9. W. F. Denning, "The Relative Powers of Large and Small Telescopes in Showing Planetary Detail," *Nature*, 52 (1895), 232.

10. Ibid., p. 233.

11. Ibid., p. 234.

12. Ibid., p. 233.

13. A. S. Williams, "On the Drift of the Surface Material in Different Latitudes," *Monthly Notices of the Royal Astronomical Society*, 56 (1896),

143. Incidentally, the nomenclature of the Jovian belts and zones that is now generally followed is also due to him.

14. A. S. Williams, "On the Rotation of Saturn," *Monthly Notices of the Royal Astronomical Society,* 54 (1893), 297.

15. E. E. Barnard, "Micrometrical Measures of the Ball and Ring System of the Planet Saturn, and Measures of the Diameter of his Satellite Titan," *Monthly Notices of the Royal Astronomical Society,* 55 (1894), 368–369.

16. Ibid., p. 370.

17. Denning, "Relative Powers of Large and Small Telescopes," p. 234.

18. E. E. Barnard, "Ball and Ring System of Saturn and Titan," p. 165.

19. Ibid., pp. 165–166.

20. Ibid., pp. 166–167.

21. Percival Lowell, "Atmosphere: In Its Effect on Astronomical Research," unpublished MS from ca. 1897, Lowell Obs. Archives.

22. James H. Worthington, "Notes on Some Foreign Observatories," *Journal of the BAA,* 12 (1912), 314; Barnard notebook in Yerkes Obs. Archives, notes dated July 2 and Sept. 17, 1907.

23. P. Lowell, "The Means, Methods and Mistakes in the Study of Planetary Evolution," unpublished MS, 1905, Lowell Obs. Archives; E. E. Barnard, E. C. Slipher, et al., "Planetary Observation," document ca. 1907 in Yerkes Obs. Archives.

24. E. M. Antoniadi, "Corpuscules en Dehors du Plan de L'Anneau de Saturn," *Bull. Soc. Ast. de France,* 23 (1909), 450.

25. W. F. Denning, "The Physical Appearance of Mars in 1886," *Nature,* 34 (1886), 105.

26. P. Lowell, "On the Kind of Eye Needed for the Detection of Planetary Detail," *Pop. Ast.,* 13 (1905), 92.

27. Worthington, "Notes on Some Foreign Observatories," p. 314.

28. P. Lowell, "Eye Needed for Planetary Detail," p. 92.

29. P. Lowell, "Means, Methods, and Mistakes."

30. P. Lowell, "Eye Needed for Planetary Detail," pp. 93–94.

31. P. Lowell, "Means, Methods and Mistakes."

32. Webb, *Celestial Objects,* vol. 1, p. 69n.

33. Ibid., pp. xxiv–xxv.

34. See, for instance, the comments of A. E. Douglass, "Atmosphere, Telescope, and Observer," *Pop. Ast.,* 5 (1897), 83–84.

35. William Noble, "Anomalies of Vision," *Astronomical Register,* 15 (1877), 222–225.

36. For Lowell, see E. P. Martz, Jr., "William Henry Pickering, 1858–1938: An Appreciation," *Pop. Ast.,* 46 (1938), 299. For Denning, see "How Many Pleiads?" *Sky and Telescope,* 70 (1985), 464–465, where, incidentally, an even more exceptional case is mentioned—that of Stephen James O'Meara, whose extraordinary acuity in the detection of planetary detail seems rivaled only by his sensitivity for faint stars (he has counted 17 Pleiads from the middle of Cambridge, Mass.).

37. William Sheehan, "On an Observation of Saturn: The Eye and the Astronomical Observer," *Journal of the Association of Lunar and Planetary*

*Observers,* 28 (1980), 150–154. What Lowell called "acuity" and "sensitivity" have nothing to do with the size of retinal cones per se but are dependent on entirely different types of receptors (the cones for acute, the rods for sensitive vision). The same eye may have excellent receptors of both types. Thus there is no reason to believe Lowell's claim that the two qualities are mutually exclusive.

38. E. M. Antoniadi, "Mars Section, Fifth Interim Report, 1909," *Journal of the BAA,* 20 (1909), 138.

## Chapter 12.
## *"The Most Brilliant Man in Boston"*

1. Ferris Greenslet, *The Lowells and Their Seven Worlds* (Boston, 1946), p. 366. Together with Abbott Lawrence Lowell's *Biography of Percival Lowell* (New York, 1935), this is the best source for Lowell's early life. For Lowell's astronomical career, the reader will find indispensable the extraordinarily detailed and well-documented studies by William Graves Hoyt, *Lowell and Mars* (Tucson, 1976) and *Planets X and Pluto* (Tucson, 1980).

2. Charles W. Eliot to Edward C. Pickering, November 22, 1894, quoted in Jones and Boyd, *Harvard College Observatory,* p. 473.

3. Nathan Appleton, *The Introduction of the Power Loom and the Origin of Lowell* (Boston, 1858), p. 9.

4. A. L. Lowell, *Percival Lowell,* p. 1.

5. Ibid., p. 5.

6. Greenslet, *The Lowells,* p. 320.

7. Quoted in Leonard, *Percival Lowell,* p. 25.

8. A. L. Lowell, *Percival Lowell,* p. 6.

9. P. Lowell, "Reply to Newcomb," *Astrophysical Journal,* 26 (1907), 131.

10. Greenslet, *The Lowells,* p. 348.

11. P. Lowell to Katharine Bigelow Lowell, October 7, 1883, quoted in A. L. Lowell, *Percival Lowell,* p. 31.

12. A. L. Lowell, *Percival Lowell,* p. 12.

13. Greenslet, *The Lowells,* p. 349.

14. Ibid.

15. P. Lowell, *Soul of the Far East,* p. 8.

16. Ibid.

17. P. Lowell to Katharine Bigelow Lowell, June 8, 1883, quoted in A. L. Lowell, *Percival Lowell,* p. 9.

18. On the other hand, Lowell's view of the oriental character has been seconded since. Kensaku Shirai, in an editorial entitled "Education in Japan: A Loss of Brilliance," writes a century after Lowell: "The stress of conformity, to the detriment of individualism, fits Japanese society.

This is a land that operates on the basis of consensus. . . . The accent is on the group. Collective accomplishment tends to take precedence over independent innovation. . . . All this has its roots in the education system, where young Japanese learn that it is better to fit in than to stand out. Or, as an old Japanese adage puts it: 'The exposed nail is beaten down'" (*Minneapolis Star and Tribune*, Aug. 3, 1983).

19. P. Lowell, *Soul of the Far East*, pp. 18–19, 21.

20. Ibid., p. 22.

21. Ibid., p. 25.

22. P. Lowell, *Mars*, pp. 141–142.

23. P. Lowell, *Soul of the Far East*, p. 207.

24. Quoted in *The Norton Anthology of English Literature*, 3d ed., ed. M. H. Abrams (New York, 1974), vol. 2, p. 307n.

25. P. Lowell, *Mars and Its Canals*, p. 11.

26. Ibid., p. 7.

27. A. L. Lowell, *Percival Lowell*, p. 60; also R. S. Dugan, "Percival Lowell," in *Dictionary of American Biography* (New York, 1933), vol. 11, p. 469.

28. P. Lowell, "Schiaparelli," *Pop. Ast.*, 18 (1910), 466.

29. Robert G. Aitken, "The Orbit of β Delphini," *Pop. Ast.*, 11 (1903), 33–34.

30. E. M. Antoniadi, "Considerations on the Physical Appearance of the Planet Mars," *Pop. Ast.*, 21 (1913), 419.

31. Webb, "Planets of the Season: Mars," p. 213.

32. Maunder, "The Canals of Mars" (1894), p. 251.

33. As noted in Antoniadi, "Physical Appearance of Mars," p. 419.

34. G. V. Schiaparelli to E. M. Antoniadi, August 29, 1909, quoted in E. M. Antoniadi, *The Planet Mars*, trans. Patrick Moore, in *Astronomy and Space* (New York, 1973), vol. 2, pp. 263–264.

35. Camille Flammarion, "Percival Lowell," *Bull. Soc. Astronomique de France*, 30 (1916), 423.

36. As noted in Emile Touchet, "La Vie et L'Oeuvre de Camille Flammarion," *Bull. Soc. Astronomique de France*, 39 (1925), 341–365. Touchet is followed here as the main source for Flammarion's life.

37. C. Flammarion, *La Planète Mars*, p. 592. This and the following translations are by Antonie Pannekoek, *A History of Astronomy* (New York, 1961), p. 378.

38. Ibid., p. 591.

39. Ibid., p. 592.

40. Flammarion's views from 1879 are found in MacPherson, *Makers of Astronomy*, pp. 196–197. Summarizing his observations of the lunar walled plain Plato, whose floor appeared to darken under more direct illumination, Flammarion wrote that this "seems opposed to all imaginable optical effect. . . . The odds are 99 to 1 that it is not the light which produces this effect, and that it is the solar heat which we do not sufficiently take into account when we are considering the modification of tints observed on the moon. . . . It is highly probable that this periodical

change of tint on the circular plain of Plato . . . is due to a modification of a vegetable nature." For W. H. Pickering's views, see, for instance, W. H. Pickering, "Lunar Changes," in "Seventh Report of Lunar Section," *Memoirs of the BAA*, 20 (1916), 110–111. The effect to which Flammarion refers is wholly illusory. In fact, the floor of Pluto does brighten under more direct illumination, as the photometer has now conclusively shown. It only appears to darken because of contrast with the dazzling regions that surround it.

41. Flammarion, *La Planète Mars*, p. 592, trans. by Pannekoek.

42. Quoted in A. L. Lowell, *Percival Lowell*, p. 93.

43. C. Flammarion, "L'Astronome Schiaparelli," *Bull. Soc. Astronomique de France*, 24 (1910), 377.

44. Sigmund Freud, *An Autobiographical Study*, trans. James Strachey (New York, 1952), p. 30.

45. P. Lowell to Frederic J. Stimson, quoted in Greenslet, *The Lowells*, p. 350.

46. A. L. Lowell, *Percival Lowell*, pp. 55–56.

47. C. Flammarion, *Death and Its Mystery*, trans. E. S. Brooks from 1902 French edition (New York, 1921), vol. 1, p. 378.

48. W. H. Pickering, "The Mountain Station of the Harvard College Observatory," *Astronomy and Astro-Physics*, 5 (1892), 353.

49. Ibid.

50. Ibid. For Douglass's life and scientific career, see George Ernest Webb, *Tree Rings and Telescopes* (Tucson, 1983).

51. W. H. Pickering, "The Mountain Station of the Harvard College Observatory," p. 355.

52. Pickering's scale assigned a value of 0 to a diffraction image consisting of a very confused, turbulent blob, 10 to a disk perfectly defined with rings motionless. A value of 6 on the scale corresponds to rings "broken but traceable." See "A Standard Scale for Telescopic Observations" (1902), in Hoyt, *Lowell and Mars*, pp. 315–318.

53. Quoted in E. S. Holden, "The Lowell Observatory, in Arizona," *Publications of the Astronomical Society of the Pacific*, 6 (1894), 165.

54. Ibid.

55. Jones and Boyd, *Harvard College Observatory*, p. 307.

56. Ibid., p. 320.

57. A. L. Lowell, *Percival Lowell*, p. 60.

58. Ibid. pp. 68–69, quoting from an unpublished introduction Lowell apparently intended for the first volume of the Lowell Observatory *Annals*.

59. P. Lowell to A. E. Douglass, Apr. 10, 1894, Lowell Obs. Archives. The account I am following here is that of Hoyt, *Lowell and Mars*, pp. 31–43.

60. P. Lowell to A. E. Douglass, Apr. 16, 1894, Lowell Obs. Archives.

61. A. E. Douglass to W. H. Pickering, Jan. 9, 1901, Lowell Obs. Archives.

62. Clyde W. Tombaugh to William Sheehan, Dec. 5, 1986.

Chapter 13.
The Visions of Sir Percival

299

Notes to
Pages
176–192

1. G. V. Schiaparelli, "Il Pianeta Marte," *Opere*, vol. 2, pp. 47–74; the third and fourth parts of this four-part essay appeared as "The Planet Mars," trans. W. H. Pickering, *Astronomy and Astro-Physics*, 13 (1894), 635–640, 714–723.

2. Schiaparelli, "The Planet Mars," pp. 635–636.

3. Ibid., pp. 638–639.

4. G. Johnstone Stoney, "Of Atmospheres upon Planets and Satellites," *Royal Dublin Society Scientific Transactions*, 6 (1898), 305–328.

5. W. W. Campbell, "The Spectrum of Mars," *Publications of the Astronomical Society of the Pacific*, 6 (1894), 228–236.

6. Alfred Russel Wallace, *Is Mars Habitable?* (New York, 1907), pp. 34–35.

7. Schiaparelli, "Planet Mars," p. 719.

8. Ibid., p. 71.

9. P. Lowell to Katharine Bigelow Lowell, May 28, 1894, quoted in A. L. Lowell, *Percival Lowell*, p. 72.

10. It has since been moved to Mount John University in New Zealand.

11. Leonard, *Percival Lowell*, p. 38.

12. Observing notebook in Lowell Obs. Archives.

13. P. Lowell, *Mars*, p. 170.

14. W. W. Campbell, review of *Mars*, by Percival Lowell, *Publications of the Astronomical Society of the Pacific*, 8 (1896), 217–218.

15. Schiaparelli, "Observations of the Planet Mars," p. 251.

16. See, for instance, P. Lowell, *Mars and Its Canals*, pp. 374–375, and Alfred Russel Wallace's trenchant critique of Lowell's argument in *Is Mars Habitable?* pp. 28–30.

17. P. Lowell, *Mars*, p. 201.

18. Ibid., pp. 167–168.

19. Greenslet, *The Lowells*, p. 366.

20. P. Lowell, *Mars*, pp. 206–207.

21. Ibid., pp. 208–209.

22. Leonard, *Percival Lowell*, p. 25.

23. Campbell, review of *Mars*, p. 215.

24. Ibid., p. 208.

25. Agnes M. Clerke, "New Views About Mars," *Edinburgh Review*, 184 (1896), 368–385.

26. G. V. Schiaparelli, "La Vita sul Pianeta Marte," *Natura ed Arte*, 4 (1895), reprinted in *Opere*, vol. 2, pp. 83–95.

27. Schiaparelli, "La Vita sul Pianeta Marte," p. 87.

28. Ibid.

29. Ibid., p. 89.

30. Ibid., p. 88.

31. Ibid., p. 91.

32. Ibid., p. 95.

33. Percival Lowell, "Mars," Lowell Obs. Archives.

34. Antoniadi, "Physical Appearance of Mars," p. 417.

35. Schiaparelli, "La Vita sul Pianeta Marte," p. 95: "Io scendo dall'Ippogrifo." His comment to Flammarion was published in a note to Flammarion's translation of "La Vita sul Pianeta Marte," which was published as "La Vie sur la Planète Mars," in *Bull. Soc. Astronomique de France,* 12 (1898), 429.

36. As quoted in MacPherson, *Makers of Astronomy,* pp. 192–193.

37. Schiaparelli, "Observations of the Planet Mars," p. 249.

38. Schiaparelli's very last pronouncement on Mars may have been in a letter dated May 19, 1910, to Franz H. Babinger, who had asked him to comment on the idea of Nobel laureate Svante Arrhenius that the darkening of the canals could be explained if they were cracks filled with hydroscopic salts, which would deepen in color as they absorbed water released from the polar caps. In his reply Schiaparelli expresses the opinion that the phenomena of Mars are "perhaps more complicated than Dr. Arrhenius believes," and that some of them seem to require the existence of "organic events of large magnitude, like the flowering of the steppes on Earth." But as for the canals, he admits that their very existence is still denied by many people, and that they "do not yet teach us anything about the existence of intelligent beings on this planet." Yet he adds, in his usual manner, "I think it would be worthwhile if somebody collected everything . . . that can reasonably be said in favor of their existence." Willy Ley, *Watchers of the Skies,* pp. 299–300.

39. P. Lowell, *Mars,* p. 150.

40. Quoted in MacPherson, *Makers of Astronomy,* pp. 192–193.

41. Maunder, "Canals of Mars," p. 249.

42. E. M. Antoniadi, "Mars Section, Fifth Interim Report, 1909," *Journal of the BAA,* 20 (1909), 138.

43. "Report of the Ordinary General Meeting of the Association, Held on March 28, 1928," *Journal of the BAA,* 38 (1928), 168.

44. Sources for the life of Maunder include H. P. Hollis, obit. notice for E. W. Maunder, *Journal of the BAA,* 38 (1928), 229–233; and E. M. Antoniadi, "E. W. Maunder," *Bull. Soc. Astronomique de France,* 42 (1928), 240–242.

45. E. Walter Maunder, "The Tenuity of the Sun's Surroundings," *Knowledge,* Mar. 1, 1894, p. 49.

46. Maunder, "Canals of Mars" (1894), p. 251.

47. Beer and Mädler, *Der Mond,* p. 129.

48. Maunder, "Canals of Mars" (1894), p. 251.

49. Ibid., p. 252.

50. Ibid., pp. 251–252.

51. Quoted in Hollis, obit. notice for E. W. Maunder, p. 232.

52. Alfred Russel Wallace, *Man's Place in the Universe* (London, 1903), p. 308.

53. Hoyt, *Lowell and Mars,* p. 201.

*Chapter 14.*
*A Stately Pleasure Dome*

*301*

*Notes to*
*Pages*
*199–204*

1. A. L. Lowell, *Percival Lowell*, p. 66, quoting from an unpublished paper written by Percival Lowell in 1897 and evidently intended as the introduction to the first volume of the observatory's *Annals.*

2. Quoted in Hoyt, *Lowell and Mars*, p. 51.

3. P. Lowell, *Mars*, p. 217.

4. P. Lowell to W. A. Cogshall, September 23, 1903, Lowell Obs. Archives.

5. P. Lowell, *Mars*, p. 139.

6. As noted in Hoyt, *Lowell and Mars*, p. 52.

7. P. Lowell, "The Means, Methods and Mistakes in the Study of Planetary Evolution," Lowell Obs. Archives.

8. E. M. Antoniadi, "On Some Objections to the Reality of Prof. Lowell's Canal System," *Journal of the BAA*, 20 (1910), 196. See also his "Further Objections," ibid., 20 (1910), 374–377.

9. Some of the notes from Lowell's log book in the Lowell Obs. Archives may be of interest:

> August 21. Venus, at last. . . . Telescope struck dome.
> August 23. Thought there might be markings with the comet-seeker.
> August 24. 2$^h$52$^m$. Plenty of markings on disk but none sure.
> > 3$^h$13$^m$. There seem to be certain markings—one black spot in special. . . .
> > 5$^h$ Apparently many markings but illusive. . . .
> August 31. First really good i.e. certain view of markings other than the spot. . . .
> September 11. Markings uncertain. . . .
> September 15. Seeing good only at times. Markings very faint and somewhat indefinite.

10. T. J. J. See, "Redécouverte et mesure du Compagnon de Sirius, faites a l'Equatorial de 24 pouces de l'Observatoire Lowell," *Bull. Soc. Astronomique de France*, 11 (1897), 160–162. See also the note on p. 276 in this journal, which recounts Schaeberle and Aitken's detection. More on See's controversial career can be found in John Lankford, "T.J.J. See's Observations of Craters of Mercury," *Journal of the History of Astronomy*, 2 (1980), 129.

11. Clyde Tombaugh to William Sheehan, Dec. 5, 1986.

12. E. M. Antoniadi, "Les Grands Instruments," *Bull. Soc. Astronomique de France*, 41 (1927), 146.

13. P. Lowell, "Markings in the Syrtis Major," *Pop. Ast.*, 4 (1896), 289–290.

14. P. Lowell, "The Rotation Period of Venus," *Astronomische Nachrichten*, no. 3406 (1896). See also his "Detection of Venus' Rotation Period and of the Fundamental Physical Features of the Planet," *Pop. Ast.*, 4 (1896), 281; and "Determination of the Rotation and Surface

Character of the Planet Venus," *Monthly Notices of the Royal Astronomical Society,* 57 (1897), 148.

15. P. Lowell, "Atmosphere: In Its Effect on Astronomical Research," ca. 1897, Lowell Obs. Archives.

16. P. Lowell, "Venus, 1903," *Pop. Ast.,* 12 (1904), 189.

17. E. M. Antoniadi, "On the Rotation Period of Venus," *Journal of the BAA,* 8 (1897), 43–46.

18. A. E. Douglass, "Atmosphere, Telescope, and Observer," *Pop. Ast.,* 5 (1897), 81.

19. Antoniadi, "Rotation Period of Venus," p. 46.

20. E. S. Holden, "Mr. Lowell's Observations of Mercury and Venus," *Publications of the Astronomical Society of the Pacific,* 9 (1897), 92.

21. Quoted in H. MacEwen, "Canals of Venus," *Journal of the BAA,* 7 (1897), 461. Schiaparelli was also supportive of Lowell's observations and considered them "the final seal of certainty" for his 225-day rotation period.

22. Dollfus's drawings are reproduced in *Planets and Satellites,* ed. G. P. Kuiper and B. M. Middlehurst (Chicago, 1961), chap. 9, pl. 5, and chap. 15, pls. 6, 7. The resemblance of some of these drawings to Lowell's is striking. Moreover, Dollfus also concluded that the spokelike markings are permanent.

In photographs, the clouds of Venus appear quite bland in visual wavelengths but show dark markings in photographs taken in the ultraviolet. It is known that individuals differ in their sensitivity to ultraviolet, and it has been suggested that this would account for differences in what various observers see on the planet. Perhaps Lowell had ultraviolet-sensitive eyes. Unfortunately, comparisons of Lowell's drawings with ultraviolet photographic markings on Venus have shown only a "general dissimilarity." See J. Caldwell, "Retrograde Rotation of the Upper Atmosphere of Venus," *Icarus,* 17 (1972), 616.

23. A. E. Douglass, "The Markings on Venus," *Monthly Notices of the Royal Astronomical Society,* 58 (1898), 383. Douglass had first reported the linear markings on Jupiter's satellites in *Astronomische Nachrichten,* no. 3432 (1897).

24. E. E. Barnard, "On the Third and Fourth Satellites of Jupiter," *Astronomische Nachrichten,* no. 3453 (1897).

25. Ibid.

26. A. L. Lowell, *Percival Lowell,* p. 98.

27. P. Lowell to A. E. Douglass, July 19, 1897, Lowell Obs. Archives.

28. See G. M. Beard, *A Practical Treatise on Nervous Exhaustion (Neurasthenia)* (New York, 1880); *American Nervousness* (New York, 1881); and *Sexual Neurasthenia* (New York, 1884).

29. Freud was one of the first to abandon neurasthenia as a diagnosis, writing in 1895: "It is difficult to make any statement of general validity about neurasthenia, so long as we use that name to cover all the things which Beard has included under it." See Sigmund Freud, "On the Grounds for Detaching a Particular Syndrome from Neurasthenia under the Description 'Anxiety Neurosis,'" in *The Standard Edition of the Com-*

*plete Psychological Works of Sigmund Freud,* trans. James Strachey (London, 1962), vol. 3, p. 90.

30. George Frederick Drinka, *The Birth of Neurosis* (New York, 1984), p. 213.

31. P. Lowell to A. E. Douglass, Apr. 21, 1897, Lowell Obs. Archives.

32. William James, "Vacations," *Nation,* 16 (1873), 90–91.

33. Quoted in Leon Edel, ed., *The Diary of Alice James* (New York, 1964), p. 149.

34. Charles K. Hofling, "Percival Lowell and the Canals of Mars," *British Journal of Medical Psychology,* 37 (1964), 33–42. Hofling says in part: "Although their surface relations were correct, there seems no question but that Lowell feared his father. It was under the latter's influence that he entered the family business upon graduation from college, managing trust funds and serving as treasurer to a textile firm. . . . It is' difficult to imagine enterprises less gratifying to a young man of Lowell's temperament. . . . It was six years until he could free himself."

35. Percival Lowell, "Comets," text dated March 10, 1910, of an article for the Boston Society of Arts, quoted in Hoyt, *Lowell and Mars,* p. 15.

36. James Russell Lowell, "The Cathedral," lines 376–379, 384–385. The poem was first published in *The Atlantic Monthly* for January 1870.

37. A. L. Lowell, *Percival Lowell,* p. 11.

38. P. Lowell, *Soul of the Far East,* pp. 22–23.

39. Freud, *Future of an Illusion,* p. 31.

40. Ibid.

41. P. Lowell, *Mars and Its Canals,* p. 383.

42. Ibid.

43. A. L. Lowell, *Percival Lowell,* p. 30; Ibid., p. 12.

44. P. Lowell, *Soul of the Far East,* p. 209.

45. A. E. Douglass to William Lowell Putnam, Mar. 12, 1901, Lowell Obs. Archives.

46. Quoted in Greenslet, *The Lowells,* pp. 354–355.

47. A. E. Douglass, "The Markings on Venus," *Monthly Notices of the Royal Astronomical Society,* 58 (1898), 383.

48. James, *Principles of Psychology,* vol. 2, p. 598.

49. Leon Daudet, *Memoirs,* quoted in Henri F. Ellenberger, *The Discovery of the Unconscious* (New York, 1970), p. 92.

50. Ellenberger, *Discovery of the Unconscious,* p. 98.

51. P. Lowell to Katharine Bigelow Lowell, June 11, 1883, quoted in A. L. Lowell, *Percival Lowell,* p. 9.

52. A. E. Douglass to W. H. Pickering, Mar. 8, 1901, Lowell Obs. Archives.

53. Quoted in Leonard, *Percival Lowell,* p. 111.

54. James H. Worthington, "Notes on Some Foreign Observatories," *Journal of the BAA,* 12 (1912), 319.

55. Quoted in H. MacEwen, "Report on Prof. Pickering's Rotation of Venus," *Journal of the BAA,* 31 (1921), 220.

56. Quoted in A. L. Lowell, *Percival Lowell,* p. 12.

57. W. Jackson Bate, *Coleridge* (New York, 1968), p. 80.

1. A. E. Douglass, "Atmosphere, Telescope, and Observer," *Pop. Ast.*, 5 (1897), 67.

2. Ibid.

3. Ibid., p. 78.

4. "Report of the Meeting of the Association, Held on March 30, 1910," *Journal of the BAA*, 20 (1910), 289.

5. Ibid.

6. Clyde Tombaugh to William Sheehan, December 5, 1986.

7. P. Lowell, "Atmosphere: In Its Effect on Astronomical Research," ca. 1897, Lowell Obs. Archives.

8. E. E. Barnard to A. E. Douglass, May 5, 1898, Lowell Obs. Archives.

9. A. E. Douglass to Joseph Jastrow, January 9, 1901, Lowell Obs. Archives.

10. P. Lowell, *Mars*, p. 159.

11. P. Lowell to A. E. Douglass, Jan. 5, 1895, Lowell Obs. Archives.

12. P. Lowell to Augustus Lowell, April 3, 1900, as noted in A. L. Lowell, *Percival Lowell*, p. 101.

13. A. E. Douglass to Joseph Jastrow, Jan. 9, 1901, Lowell Obs. Archives.

14. P. Lowell, "The Markings on Venus," *Astronomische Nachrichten*, no. 3823 (1902).

15. "Report of the Annual Meeting of the Association, Held on Oct. 29, 1902," *Journal of the BAA*, 13 (1903), 13.

16. P. Lowell, "Venus, 1903," p. 184.

17. Ibid.

18. E. Walter Maunder and J. E. Evans, "Experiments as to the Actuality of the 'Canals' of Mars," *Monthly Notices of the Royal Astronomical Society*, 63 (1903), 498.

19. Read by Maunder at the meeting of the BAA, Dec. 30, 1903, and reported in "Report of the Meeting of the Association Held on Dec. 30, 1903," *Journal of the BAA*, 14 (1904), 118.

20. Vincenzo Cerulli, *Marte nel, 1896–97* (Collurania, It., 1898), p. 105, quoted in E. M. Antoniadi, "Mars Section, Fifth Interim Report, 1909," *Journal of the BAA*, 20 (1909), 139.

21. Quoted in E. M. Antoniadi, "Report of Mars Section, 1903," *Memoirs of the BAA*, 16 (1910), 58.

22. P. Lowell, "The Canals of Mars—Photographed," *Pop. Ast.*, 13 (1905), 479.

23. Ibid., p. 481.

24. P. Lowell, *Mars and Its Canals*, p. 277. Schiaparelli, ever the master of ambiguity, exclaimed: "I should never have believed it possible!" Quoted in P. Lowell, *Mars as the Abode of Life*, p. 155.

25. W. T. Lynn, *Journal of the BAA*, 16 (1906), 162.

26. For the life of Eugéne Marie Antoniadi, see obituary notice by F. Baldet, *Bull. Soc. Astronomique de France*, 58 (1944), 58–60; also obit. in *Journal of the BAA*, 55 (1945), 163–165.

27. E. M. Antoniadi, "Report of Mars Section, 1896–97," *Memoirs of the BAA*, 5 (1897), 75–122.

28. E. M. Antoniadi, "Report of Mars Section, 1898–99," *Memoirs of the BAA*, 9 (1901), 68.

29. P. Lowell, "Venus, 1903," p. 189.

30. P. Lowell, "The Means, Methods and Mistakes."

31. E. M. Antoniadi, "Report of the Mars Section, 1900–01," *Memoirs of the BAA*, 11 (1903), 137.

32. E. Walter Maunder, "A New Chart of Mars," *Observatory*, 26 (1903), 351.

33. Ibid., p. 354.

34. E. M. Antoniadi to P. Lowell, Sept. 9, 1909, Lowell Obs. Archives.

35. P. Lowell to E. M. Antoniadi, Sept. 26, 1909, Lowell Obs. Archives.

36. See particularly Clyde W. Tombaugh and Patrick Moore, *Out of the Darkness: The Planet Pluto* (New York, 1980), pp. 80–81.

37. E. M. Antoniadi, "Report of Mars Section, 1909," *Memoirs of the BAA*, 20 (1916), 28.

38. E. M. Antoniadi, "Mars Section, Fourth Interim Report, 1909," *Journal of the BAA*, 20 (1909), 78.

39. E. M. Antoniadi to P. Lowell, Oct. 9, 1909, Lowell Obs. Archives.

40. P. Lowell to E. M. Antoniadi, Nov. 2, 1909, Lowell Obs. Archives.

41. E. M. Antoniadi to P. Lowell, Nov. 15, 1909, Lowell Obs. Archives.

41. E. M. Antoniadi to P. Lowell, Nov. 15, 1909, Lowell Obs. Archives.

42. E. M. Antoniadi, "Mars Section, Fourth Interim Report, 1909," p. 79.

43. When I first came across this little sketch in the Lowell Observatory archives in 1982, I experienced a remarkable sense of déjà vu; I had seen such detail before—on the Mariner and Viking spacecraft maps. It appears in the letter of E. M. Antoniadi to P. Lowell, Nov. 15, 1909, in the Lowell Obs. Archives.

44. E. M. Antoniadi to P. Lowell, Oct. 9, 1909, Lowell Obs. Archives.

45. Antoniadi, "Mars Section, Fourth Interim Report, 1909," p. 79.

46. Ibid.

47. E. M. Antoniadi, "Report of Mars Section, 1909," *Memoirs of the BAA*, 20 (1916), 32.

48. Antoniadi's first notice of his observations of September 20 had been sent to an Athenian publication and published on September 28, 1909. Ironically, Frost's telegram appears not to have referred to any particularly detailed observations with the 40-inch but to have been composed as an intentionally ambiguous response, for of course Lowell would have agreed with the purport of it just as Antoniadi did, but for

opposite reasons. See Edwin B. Frost, *An Astronomer's Life* (Boston and New York, 1933), pp. 217–218.

49. Quoted in François Terby, *Areographie, Mémoire Académie de Belgique, Savants Etranger,* 39 (1874), 55.

50. As noted in E. M. Antoniadi, "Mars Section, Fifth Interim Report, 1909," p. 138.

51. G. E. Hale to E. M. Antoniadi, Jan. 3, 1910, quoted in E. M. Antoniadi, "Mars Section, Sixth Interim Report, 1909," *Journal of the BAA,* 20 (1910), 191–192. Hale used 800× on the reflector "to show the finest details" but found "no trace of straight lines, or geometrical structure." He noted that "a few of the larger 'canals' of Schiaparelli were seen, but these were neither narrow nor straight. On one occasion I could see the two 'canals' which reach out from the extremities of *Sabaeus Sinus* resolved into minute curved and twisted filaments" (pp. 191–192).

52. E. M. Antoniadi, "On Some Objections to the Reality of Prof. Lowell's Canal System of Mars," *Journal of the BAA,* 20 (1910), 196.

53. Antoniadi, "Mars Section, Sixth Interim Report, 1909," p. 189.

54. "Report of the Meeting of the Association March 30, 1910," pp. 288–289. Lowell had sent his *Nature* essay to Antoniadi in January and had sent an additional note a few days later: "I am sending for a paper recently delivered on Venus [*sic*] in which you will see the strength of the evidence [concerning the markings of that planet] and I enclose with this an explanation of why the lines on Mars might show as a mosaic in a large glass. If you will test it on a star you will see its force."

55. E. M. Antoniadi, "Further Objections to Prof. Lowell's Canal System of Mars," *Journal of the BAA,* 20 (1910), 374–377.

56. Sir Isaac Newton, *Opticks,* 4th ed. (London, 1730), bk. 1, pt. 1, prop. 8, prob. 2.

57. E. E. Barnard to E. M. Antoniadi, May 27, 1910, quoted in Antoniadi, *The Planet Mars,* p. 255.

58. Antoniadi, "Physical Appearance of Mars," pp. 418–419.

59. "Report of the Meeting of the Association, March 30, 1910," p. 287.

60. Antoniadi, "Le Retour de la Planète Mars," pp. 348–349.

61. Antoniadi, "Physical Appearance of Mars," p. 420.

62. Antoniadi, "Mars Section, Sixth Interim Report, 1909," p. 191.

63. Antoniadi, "Le Retour de la Planète Mars," p. 350.

64. Antoniadi, "Les Grands Instruments," p. 146.

65. "Report of the Meeting of the Association, March 30, 1910," p. 289.

66. Ibid., p. 291.

67. M.E.J Gheury, "Prof. Lowell's Address on Mars," *Journal of the BAA,* 20 (1910), 385–386.

68. Gheury, "Lowell's Address on Mars," p. 385. The physicists in question were H. Pender of Johns Hopkins University and V. Cremien of Paris, who were attempting to experimentally prove the existence of a magnetic field around a moving electrically charged body.

69. Rev. T.E.R. Phillips, "A Visit to the Meudon Observatory: Jupiter in a Great Telescope," *Observatory,* 34 (1911), 365–366.

70. Quoted in Leonard, *Percival Lowell,* p. 29.

71. Ibid.

72. P. Lowell, "Schiaparelli," *Pop. Ast.,* 18 (1910), 456.

73. See Hoyt, *Planets X and Pluto,* esp. pp. 83–141.

74. P. Lowell, "Memoir on Saturn's Rings," *Memoirs of the Lowell Observatory,* vol. 1, no. 2 (Sept. 7, 1915). Lowell says in part: "Long ago Kirkwood explained Cassini's division as due to perturbation by Mimas. . . . That the rings have thus been sculpted by Saturn's nearest satellite, minute as it is, subsequent investigation at this observatory has, in surprising detail, confirmed. Indeed if we compute the distance where particles of the ring system would have periods commensurate with that of Mimas and then compare the table with the system we shall be struck by the method displayed in its modelling as it appeared at the Lowell Observatory in 1915" (p. 3). For the post-Voyager view of the ring system, see Jeffrey N. Cuzzi, "Ringed Planets: Still Mysterious," *Sky and Telescope,* 67 (1984), 511–515; 68 (1985), 19–23.

75. Quoted in Hoyt, *Lowell and Mars,* p. 297.

76. P. Lowell, "Our Solar System," *Pop. Ast.,* 24 (1916), 427.

77. P. Lowell, *The Soul of the Far East,* p. 5.


*Chapter 16.*
*Planets and Perception*


1. William James, *Principles of Psychology,* vol. 2, pp. 296–297.

2. Clerke, *Popular History of Astronomy,* pp. 325–326.

3. Francis Bacon, "Of Truth," in *The Works of Francis Bacon,* ed. James Spedding and Robert Leslie Ellis (Boston, 1860), vol. 12, p. 82.

4. Carl Sagan, *Cosmos* (New York, 1980), p. 111.

5. *Boswell's Life of Johnson,* 6 vols., ed. G. B. Hill, rev. and enlarged by L. F. Powell (Oxford, 1934–1950), vol. 1, p. 40.

6. P. Lowell, "Experiment on the Visibility of Fine Lines," *Lowell Observatory Bulletin,* no. 2 (1903); and V. M. Slipher and C. O. Lampland, "Notes on Visual Experiment," *Lowell Observatory Bulletin,* no. 10 (1903). Interestingly, E. E. Barnard had performed similar experiments a number of years earlier. During the transit of Mercury of November 9–10, 1894, he had been intrigued by the fact that he was unable to see Mercury against the solar disk with the naked eye, in spite of its having an apparent diameter of 11″ of arc. On further investigation, he found that "a small wire whose diameter was 0.009 of an inch, suspended against the bright sky, was distinctly seen with the unaided eye at a distance of 356 feet. At that distance it subtended an angle of only 0″.44." See E. E. Barnard, "Micrometrical Determinations of the Dimensions of the Planets and Satellites, of the Solar System, Made with the 36-inch Refractor of the Lick Observatory," *Pop. Ast.,* 5 (1897), 286n. Compare also S. Hecht and E. U. Mintz, "The Visibility of Single Lines at Various

Illuminations and the Retinal Basis of Visual Resolution," *Journal of General Psychology*, 22 (1939), 593–612. Hecht and Mintz found that for a single dark line against a bright background the minimum attainable resolution is 0".5 of arc and that this corresponds to an intensity decrement of only 1 percent between the line and the background.

7. P. Lowell, *Mars as the Abode of Life*, p. 275.

8. Maunder, "Canals of Mars" (1894), p. 251.

9. Schiaparelli, "The Planet Mars," p. 722.

10. The figure shown was first discussed by Gaetano Kanizsa in "Contours without Gradients or Cognitive Contours?" in *Italian Journal of Psychology*, 1 (1974), 93–112.

11. Arthur A. Hoag to William Sheehan, Sept. 28, 1984.

12. Observing notebook in Lowell Obs. Archives, notes dated June 7, 9, 1894.

13. P. Lowell, *Mars*, p. 139.

14. Observing notebook in Lowell Obs. Archives, notes dated Aug. 21, 22, 1894.

15. Quoted in E. M. Antoniadi, "Report of Mars Section, 1898–1899," *Memoirs of the BAA*, 20 (1901), 68.

16. E. H. Gombrich, *Art and Illusion: A Study in the Psychology of Pictorial Representation*, rev. ed. (Princeton, N.J., 1961), pp. 204–205.

17. P. Lowell, *Mars and Its Canals*, pp. 174–175.

18. Antoniadi, *The Planet Mars*, p. 252.

19. Antoniadi, "Report of Mars Section, 1909," p. 38.

20. Quoted in Leonard, *Percival Lowell*, p. 71.

21. Quoted in E. M. Antoniadi, "Report of Mars Section, 1907," *Memoirs of the BAA*, 17 (1907), 67.

22. R. J. Trumpler, *Lick Observatory Bulletin*, no. 387 (1927), p. 39.

23. Clyde Tombaugh to William Sheehan, Dec. 5, 1986.

24. W. H. Pickering, "Recent Studies of the Martian and Lunar Canals," *Pop. Ast.*, 12 (1904), 77.

25. E. Walter Maunder and J. E. Evans, "Experiments as to the Actuality of the 'Canals' of Mars," *Monthly Notices of the Royal Astronomical Society*, 63 (1903), 498.

26. "Report of the Meeting of the Association, March 30, 1910," p. 291.

27. The basic idea of how the retina accomplishes this was worked out by the Austrian physicist Ernst Mach in his analysis of the phenomenon now known as Mach bands. His classic paper is "Ueber die Wirkung der raumlichen Verteilung des Lichtreizes auf die Netzhaut," *Sitzungs-berichte der Wiener Akademie*, 52 (1865), 303–322. For an English translation of this and other papers by Mach and a discussion of retinal neural networks generally, see Floyd Ratliff, *Mach Bands: Quantitative Studies on Neural Networks in the Retina* (San Francisco, 1965).

28. "Report of the Meeting of the Association Held Dec. 31, 1890," *Journal of the BAA*, 1 (1890), 112; Webb, "Planets of the Season: Mars," p. 213.

29. D. H. Hubel and T. N. Wiesel, "Receptive Fields of Single Neurones in the Cat's Striate Cortex," *Journal of Physiology* (London), 148 (1959), 574–591.

30. Gombrich, *Art and Illusion*, p. 73.

31. John H. Flavell and Juris Draguns, "A Microgenetic Approach to Perception and Thought," *Psychological Bulletin*, 54 (1957), 198–199.

32. Percival Lowell, "Atmosphere: In Its Effect on Astronomical Research," Lowell Obs. Archives.

33. Antoniadi, "Mars Section, Sixth Interim Report, 1909," p. 189.

34. G. V. Schiaparelli to François Terby, June 8, 1888, quoted in Maunder, "The Canals of Mars" (1888), pp. 347–348.

35. E. M. Antoniadi, "On the Advantages of Large Over Small Telescopes in Revealing Delicate Planetary Detail," *Journal of the BAA*, 21 (1911), 105; Rev. T.E.R. Phillips, "A Visit to the Meudon Observatory: Jupiter in a Great Telescope," *Observatory*, 34 (1911), 365–366.

36. See, for instance, Carl Sagan and Paul Fox, "The Canals of Mars: An Assessment after Mariner 9," *Icarus*, 25 (1975), 602; and Patrick Moore, "Requiem for the Canals," *Journal of the BAA*, 87 (1977), 589.

37. E. M. Antoniadi, "Mars Section, Third Interim Report, 1909," *Journal of the BAA*, 20 (1909), 27.

38. Antoniadi, "Mars Section, Fifth Interim Report, 1909," p. 141.

39. Antoniadi, "Physical Appearance of Mars," p. 420.

40. Maunder, "The Canals of Mars" (1894), p. 251.

# FIGURE CREDITS

Figs. 2.1–2.3. Arthur Berry, *A Short History of Astronomy: From Earliest Times Through the Nineteenth Century* (New York, 1961; reprint of 1898 edition), figs. 53, 67, 66.

Figs. 5.1, 5.2, 5.4. Camille Flammarion, *La Planète Mars, et ses conditions d'habitabilité* (Paris, 1892).

Fig. 5.3. Richard A. Proctor, *Other Worlds Than Ours* (New York, 1871), p. 104.

Figs. 6.1, 7.1. Amand Freiherr von Schweiger-Lerchenfeld, *Atlas der Himmelskunde* (Vienna, 1898), pp. 120, 185.

Fig. 7.2. Percival Lowell, *The Evolution of Worlds* (New York, 1909), opp. p. 222.

Fig. 7.3. Left: G. V. Schiaparelli, "Sulla rotazione di Mercurio," *Astronomische Nachrichten*, no. 2944 (1889); right: E. M. Antoniadi, "La Planète Mercure: Sa Géographie, sa Rotation et ses Voiles Atmosphériques," *Bulletin Société Astronomique de France*, 47 (1933), 545.

Figs. 7.4, 8.2. Flammarion, *La Planète Mars, et ses conditions d'habitabilité* (Paris, 1892).

Fig. 8.1. George Phillips Bond, "Observations on the Great Nebula of Orion," ed. Truman Henry Safford, *Annals of the Harvard College Observatory* (Cambridge, Mass., 1867), vol. 5, frontispiece.

Figs. 8.3, 8.4. N. E. Green, "Observations of Mars, at Madeira in Aug. and Sept., 1877," *Memoirs of the Royal Astronomical Society*, 44 (1877–79), plates fol. p. 138.

Fig. 10.1. T.E.R. Phillips, et al., *Splendour of the Heavens* (London, 1925), vol. 1, p. 306.

Figs. 10.2, 11.1, 11.2. Yerkes Observatory photographs, University of Chicago, Williams Bay, Wisconsin.

Fig. 12.1. Courtesy of Harvard University Archives.

Figs. 13.1, 14.1. Courtesy of Lowell Observatory, Flagstaff, Arizona.

Fig. 14.2. Percival Lowell, "Determination of the Rotation Period and Surface Character of the Planet Venus," *Monthly Notices of the Royal Astronomical Society*, 57 (1897), pl. 5.

Fig. 14.3. Photograph by author.

Fig. 15.1. E. M. Antoniadi, "Mars Section, Fourth Interim Report, 1909," *Journal of the British Astronomical Association*, 20 (1909), pl. 1.

Fig. 15.2. Courtesy of Lowell Observatory, Flagstaff, Arizona.

Fig. 15.3. T.E.R. Phillips, "Sixteenth Report of the Section for the Observation of Jupiter, Apparition of 1911," *Memoirs of the British Astronomical Association*, 19 (1916), pl. 1.

Fig. 15.4. Percival Lowell, "Memoir on Saturn's Rings," *Memoirs of the Lowell Observatory*, vol. 1, no. 2 (1915), pl. 1.

Fig. 16.2. Courtesy of Lowell Observatory, Flagstaff, Arizona.

Fig. 16.3. Courtesy of U. S. Geological Survey, Flagstaff, Arizona.

References to illustrations are printed in italics.

on, 192–193; spectroscopic observations of, 177–178; volcanoes on, 147–149, *148;* water on, 177–178, 189–193. *See also* Martian canals

*Mars* (Lowell), 110–111, 185–187, 199, 228, 265; reviews of, 110–111, 187, 189–190

*Mars and Its Canals* (Lowell), 234, 266, 268

Martian canals, *92–93, 102, 125, 148, 267*
—artificial appearance of, 101–103
—compared with other planetary markings, 208–209, 237–238
—concepts of: Antoniadi, 91, 112, 155, 247–248; Green, 169; Lowell, 5–6, 91, 94, 184–187, 189–190, 192–194, 231–235, 237, 254, 256, 260, 276; Maunder, 91, 105, 195–197, 231–232, 238–239, 249–250, 269–270, 273; Schiaparelli, 86–88, 135, 179, 190–194, 272
—diffraction analysis of, 247–248
—discovery of, 3–4, 62–63, 90–91
—experiments on, 228–229, 231–232
—first verifications of, 124, 126–131, 135
—geminations of, 103, 142, 191–192, 228
—mass hysteria and, 110
—observations by: Antoniadi, 112, 235–237, 268, 273, 276–277; Barnard, 142; Denning, 151; Lowell, 181, 265–268, *267;* Maunder, 131; Schiaparelli, 3–4, 90–93, 101–103, 105–106, 108–112, 113, 121–122, 124–131, 135, 142, 165, 244, 247–248, 272,

276, *92–93, 102, 125*
—photography of, 216, 233–234
—resolution into irregular details, 244–251
—in "seas," 187, 202
—straightness of, 91, 94, 184–185
—subjective contour explanation, 269
—tachistoscope effect and, 265–269, *267*
—visibility of, 90–91

Martians, attempts to communicate with Earth, 136. *See also* Mars, life on.

Maunder, Edward Walter, 7, 65, 84, 91, 105, 108, 131, 166, 195–198, 230–232, 238–239, 249–250, 269–270, 273, 278; *Are the Planets Inhabited?* (1903), 198; biographical sketch of, 195–196; canal observations by, 131; concepts on extraterrestrial life, 198; criticisms of Martian canals, 105, 196–197, 231–232, 238–239, 249–250, 269–270, 273; experiments on Martian canals, 231–232; minimum visibility of lines, ideas about, 195–196, 249–250, 260, 269–270; religious views of, 197–198; solar observations by, 195–196

Maxwell, James Clerk, 64, 177

Medici, Giuliano de', 15

Memory, role of, in planetary observation, 7–8

Mercury, 71–86, 150, *77, 80, 81*
—analogy: to Earth, 71, 75–76; to Moon, 73–75
—atmosphere of, 75–76, 78, 83–85
—clouds on, 75–76, 78, 83–85
—elongations of, 71–72, 82–83
—inhabitants of, 75–76

Orion Nebula, 95–98, 129, *97*
Owen-Huxley controversy, 237

Paley, William, 3–4; *Natural Theology* (1802), 3–4
*Paradise Lost* (Milton), 20, 23
Paris Observatory, 168
Pavonis Mons (Martian volcano), 147
Peirce, Benjamin, 158, 194
Perception, artificial motifs in, 269–270, 276–278; Gestalt theories of, 262, 264; microgenesis of, 271–272; tachistoscope effect and, 265–273, *267*
*Percival Lowell: An Afterglow* (Leonard), 220
Perrotin, Henri, 76, 122, 130–131, 134
Perseid meteors, 70
Phillips, Rev. T.E.R., 126, 236, 251, 253, 265–266
Phison (Martian canal), 90, 131, 163
Phobos. *See* Mars, satellites of
Photography, astronomical, 95–98, 130; of Martian canals, 216, 233–234
Piaget, Jean, 3
Pic du Midi Observatory, 79
Pickering, Edward C., 157, 171–173
Pickering, William Henry, 136, 171–174, 176, 181, 183, 196, 204, 228, 232, 269; concepts on astronomical seeing, 171–174; discovery of Martian lakes, 136, 172–173; polariscopic observations of Mars, 183. *See also* Douglass, A. E.; Harvard College Observatory, Boyden Station of Planetary observation, obstacles to, 6–7. *See also* Chromatic aberration; Diffraction; Large vs. small telescope debate; Seeing, astronomical
Planetology, 257

Planets, motion of, 10; rotation of, 71. *See also* individual names.
Plato, 101, 270
Pleiades, 14, 155
Plutarch, 11
Pluto, 257
Polar caps. *See* Mars, polar caps of
Polaris, 113–114
Polariscope, 183
*Popular History of Astronomy During the Nineteenth Century* (Clerke), 29, 136
*Practical Astronomer* (Dick), 138
Preperceiving, 63–67. *See also* Expectation, role of, in planetary observation
*Principles of Psychology* (James), 66–67, 258
Proclus, 101
Proctor, Richard Anthony, 27, 33–34, 47–48, 88, 108, 110
Procyon, 62
Psychic phenomena, 169–171, 215–216. *See also* Flammarion, C.; Lowell, P.; Schiaparelli, G. V.
Psychoanalysis, 212–213
Ptolemaic system, 22
Ptolemy, Claudius, 126
Pulkova Observatory, 58, 62, 67, 124; 15-inch Merz refractor at, 58, 62
*Purchas: His Pilgrimage* (Purchas), 163
Putnam, William Lowell, 173, 229
Pythagoreans, 10

Quételet, Lambert Adolphe Jacques, 65

Radar observations of Mercury, 79
Raffles, Rev. Thomas, 47
Red shifts of nebulas, 198
Refractors, achromatic, 44
Resolving power of telescopes, 50–54
Riccioli, Giovanni, 16, 21
Rigel, 50

*William Sheehan, an amateur astronomer since the age of nine, observes regularly with various telescopes, of which his main one is a 6-inch refractor. His special interest is in the Moon and the planets, but over the years he has become as intrigued by what happens at the eye end of the telescope—the unjustly neglected perceptual and psychological factors that enter into visual planetary observations—as by the planets themselves.*

*Dr. Sheehan earned a B.A. from the University of Minnesota, an M.A. from the University of Chicago, and an M.D. degree from the University of Minnesota. He lives in St. Paul, Minnesota, with his wife, Deborah, and their son, Brendan.*